RADIATION PROTECTION MANUAL

3rd Edition

by

LITA LEE, Ph.D.

SOURCES -FALLOUT, IRRADIATED FOOD, TV'S VDT'S, MICROWAVE OVENS... DAMAGE TO THE HUMAN BODY... THE PROTECTIVE NATURE OF MOTHER'S MILK, HOW TO PROTECT YOURSELF USING DIET AND NUTRITIONAL SUPPLEMENTS...NATURAL ANTI-CARCINOGENIC FOODS

Preface

In 1986, I wrote the first edition of this manual. At that time, information on the health effects of radiation was available but hidden from the public eye and certainly not available in the main stream media. One had to look in grassroots tabloids and self-published books and articles. When I was invited to lecture on how to protect oneself from radiation, very few people attended the lecture, as if they did not know, did not believe or did not care much about radiation as a topic of health. There were virtually no books available on natural methods of radiation protection. The following year, there was a virtual explosion of information on radiation, its effects, the cover-up of its effects by the radiation industry, and - four new books on radiation protection.

Since then, radiation - what it is, what it does, and what to do about it - has become a "hot topic" if you'll pardon the pun. We've gone from looking the other way, through denial, the "oops stage", sheepish admission of the hazards and the damage, to the "now what" stage.

I wrote this book not as a technical treatise but to offer the lay public a practical way of dealing with the radiation problem. I have touched only the surface of what radiation is and what it does. I leave the technical details to the many books and scientific papers now available, which I have referenced for those with a deeper interest. So don't expect this book to prove to you that radiation is a problem. I have merely described today's conditions as I see them .

I do this with great optimism and hope for the future of the earth and its inhabitants. Radiation is here to stay. It will not go away for as long as the earth is here. But there are natural methods of protection that you should know about, for yourself and your children. Learning about these methods of radioprotection has made a great difference in my life - the difference between living in fear and hope.

The Third Edition was re-written to reflect the importance of enzyme nutrition, and the new availability of different kinds of enzymes, which only recently have emerged after years of research in enzyme chemistry and plant genetics.

I have listed footnotes in the Reference Section. Appendix A is a list of people, organizations and products which have a healing or protective effect upon the earth and its inhabitants. This is designated by (A). Appendix B is a list of books described in the text, designated by (B).

I have listed many products and companies by names. No one is paying me to endorse their products. I am listing them only to tell you about products that I am familiar with because I use them and have observed their effects. I do not mean to exclude other similar products but can only speak from my own experience.

Products designated by (*) can only be obtained from a health practitioner. Sources for other products are listed wherever possible.

In February 1991, Grassroots Network will publish the first issue of my quarterly newsletter which will encompass many environmental problems, not just radiation, but other health concerns and what to do about them. Contact me for more information.

Yours in health,

Lita Lee

Lita Lee

Acknowledgments

I would like to express my gratitude and appreciation to all those who have contributed to this book both technically and by their encouragement.

In particular, I want to thank my husband, John Flinn, whose technical expertise and research helped clarify some of the more complex details on radiation. And for staying up with me during the wee hours as a helpful companion and friend.

Thanks and appreciation go to Master Photographer, Fred English, for his beautiful cover photo of our twins, Shawn and Veronica.

I give thanks to Ed Donnellan for his cartoons, specially created for this book, and to the other cartoonists, Tony Auth, Sidney Harris, Etta Hulme and Mr. Stayskal, for giving permission to use their cartoons in this book.

This edition would not have been possible without the editing, excellent suggestions and continuous support of my assistant, Joyce Jordan.

The following people contributed technical data, new product information, and research results to this book. They and their affiliations are listed in the appendices. Many of these people are involved in developing and marketing new products which will contribute to a safer and healthier environment: Dr. Howard Loomis, Jim Fahey, Zane Baranowski, Scott Springer, Steve Mcleod, Joel Hill, Alan Horn, Simon Martin, Dr. Khem Shahani, John Banta. and Ms. Egan O'Conner.

Printed by Spilman Printing, 1801 Ninth St., Sacramento, CA 95814.

Published by Grassroots Network, 2061 Hampton Avenue, Redwood City, CA 94061. Phone: (415) 369-2554.

To order, send $6.95 + $3 postage and handling to the publisher.

Disclaimer

Table of Contents

Table of Contents

INTRODUCTION

We are living in an age where everything we eat, breathe and drink is contaminated in one way or another. Pesticides, insecticides, herbicides, heavy metals (cadmium, mercury, lead, arsenic, etc.), depleted soils full of chemicals, polluted water, asbestos, smog, cigarette smoke, and **RADIATION** - ionizing and non-ionizing. What are their effects on humans and animals? They all produce **FREE RADICALS**. And what do free radicals produce? Cancer, aging, heart disease, and inflammatory processes such as arthritis and allergic reactions.

Inflammation we can feel. Aging we can see. But radiation is elusive. You can't see it, taste it, or smell it. Most people can't detect a dose large enough to cause cancer either. And they are still arguing about how large this dose must be.

Recent articles in *Science News* and other journals[1,2] have reported that the lethal dose of radiation for human bone marrow may be as low as 154 rads, not the previously agreed-upon 600-rad dose which is used to estimate radiation casualties in a nuclear war.

Can we build human-error-proof nuclear plants? Radiation industry experts suggest that we can. As a chemist who has spent 32 years in laboratories and chemical plants, I disagree. There wasn't one week during the entire 32 years that someone did not have an accident involving spillage, breakage, explosions, mechanical failure, human error in calculating how much to add and in adding the wrong chemicals or pushing the wrong switches, dials, etc. Even under the best safety conditions, I have witnessed terrible accidents involving irreparable human injury. It's one thing to lose your eye or be peppered with glass and flying fragments. It's quite another to pepper the earth and its inhabitants with radioactivity.

There are other sources of ionizing radiation besides nuclear plant fallout. Medical techniques that involve ionizing radiation include nuclear medicine, x-rays, fluoroscopy, mammography, and C.A.T. scans. How much is harmful? Can we decrease the x-ray dosage and still obtain useful diagnostic information? What about irradiated food and the nuclear waste crisis? Are underground nuclear weapons tests environmentally safe? Where does all that radioactivity go after the blast?

There are sources of non-ionizing radiation that most people don't think about because they are sources of convenience, education and pleasure: tanning booths, microwave ovens, video display terminals (computer monitors), color televisions, cellular telephones, and the power lines that make all this wonderful technology possible. Are these harmful? All of these topics are discussed in this manual as well as simple foods and nutritional supplements which protect the body from radiation damage and other sources of free radicals.

"We'd like to reiterate... The procedure is safe and the side effects, if any at all, are minimal..."

SECTION I:
RADIATION, WHAT IT IS, WHAT IT DOES

There are two types of radiation, ionizing and non-ionizing. What's the difference? What are the effects?

CHAPTER 1:
RADIATION AND FREE RADICALS

Ionizing Radiation

Most people are familiar with ionizing radiation such as x-rays and gamma rays which are of high frequency, penetrating and can cause fatal diseases, cancer and genetic damage. The word 'ionizing' means that this type of radiation is of sufficient energy to change atoms into charged particles called ions. These ions can be positively or negatively charged, depending on whether they loose or gain electrons, negative particles that are part of the composition of atoms. Ionizing radiation is produced naturally from cosmic rays and from nuclear bombs, nuclear reactors, medical and dental x-rays, and is the type of radiation used to irradiate food. It also comes from unstable radioactive elements which become stable by emitting energy as electromagnetic waves and high energy particles. The electromagnetic spectrum designates radiation in its various forms from low frequency, long wavelengths to high frequency short wavelengths.

Non-ionizing Radiation

Of lower energy and frequency and longer wavelength but nonetheless damaging is non-ionizing radiation which starts in the middle of the ultraviolet spectrum and, moving from higher to lower frequencies, includes sunlight, visible light, infrared light, microwaves, radar, microwave ovens, radio frequencies, television, very low and extremely low frequencies (VLF, ELF) and power lines. Video display terminals (VDT's) or computer terminals emit a wide gamut of radiation including soft x-rays (in the ultraviolet) all the way down to VLF and ELF radiation. We will talk more about this later.

Free Radicals

The common effect uniting both types of radiation is the production of free radicals, and radiation sickness is basically a free radical process. What is a free radical? It is a highly reactive species with an odd (lone or unpaired electron). This reactive species likes to initiate chain reactions wherever it is. If it's inside your body, watch out! There's a domino game going on and what's falling? Not domino pieces but your cells, enzymes and genes - the living substances composing your body. Free radicals are formed in your body:

* during normal body metabolism and during aging;
* when fats become rancid (peroxidized), especially unsaturated fats;
* when you sunbathe, breathe polluted air, drink polluted water or eat junk foods (processed non-foods devoid of nutrients and containing chemicals and/or drugs);
* during all disease processes (cancer, arteriosclerosis, diabetes, etc.);
* during all inflammatory processes (conditions ending in **"itis"** such as arthritis, bronchitis, sinusitis, colitis, cystitis)
* when you are exposed to any kind of radiation.

The bottom line in radiation-produced free radical pathology is cancer, either in the weakest organ or in the organ targeted by the radioactive isotopes you have been hearing about in the news since Chernobyl: elements such as iodine 131, cesium 137 and strontium 90.

Why does radiation produce free radicals? Because it gives the chemical bond enough energy to tear it apart into two halves, each retaining a single electron. The two new odd-electron particles are free radicals. Any form of energy sufficient to break the chemical bond can produce free radicals. You will be surprised at the list. It includes not only nuclear radiation and x-rays but all forms of non-ionizing radiation listed above, as well as chemicals and their by products (insecticides, chlorinated hydrocarbons, etc.) and minerals (iron, cadmium, etc.). So you need to know about foods and supplements which can protect you from these toxic chemicals and free-radical producing sources.

There are many kinds of free radicals but the type I am talking about here are toxic oxygen species. How, you may ask, can life-giving oxygen and water molecules (containing oxygen) become toxic? It happens when ionizing radiation knocks electrons out of their orbits and when these reactive electrons are captured by oxygen or water in our cellular fluids. Let's name some of these nasty free radicals. In order of decreasing reactivity, some of these species include: hydroxyl radical, the most energetic ($HO\cdot$), superoxide radical anion ($O_2^{\cdot-}$), singlet oxygen ($^1O_2$) and peroxide radical ($HOO\cdot$).

To make this list of free radical producers more complete, let's not forget cigarettes. Did you know they too are a source of radioactivity? Recent research at the Massachusetts Medical Center indicates that a 1-1/2-pack per day smoker gets a yearly dose of radiation in parts of his lungs equal to about 300 chest x-rays.[3] Scientists are just beginning to find out how much harm the radioactive isotopes in cigarettes do.

Many people ask, 'if radiation is natural, what is all the fuss about?' Nature could not predict the great increase in ionizing radiation that came from the growth of the nuclear industry - not just nuclear reactors and underground nuclear tests, but nuclear medicine, x-rays, ionizing smoke detectors, food and medical supply irradiators plus non-ionizing radiation sources such as TV's, VDT's, microwave ovens and other wonders of our electronic age. You can see that there is more radiation present than we are equipped to handle. Paul Brodeur emphasized the increase in non-ionizing radiation in a 1986 *US Magazine* article, "*the U.S. is polluted with levels of electromagnetic radiation - sometimes called electronic smog - which are estimated to be 100 to 200 million times as great as the natural non-ionizing background radiation. Such levels have no counterpart in history. To keep on pretending that they are harmless is ridiculous.*" 4

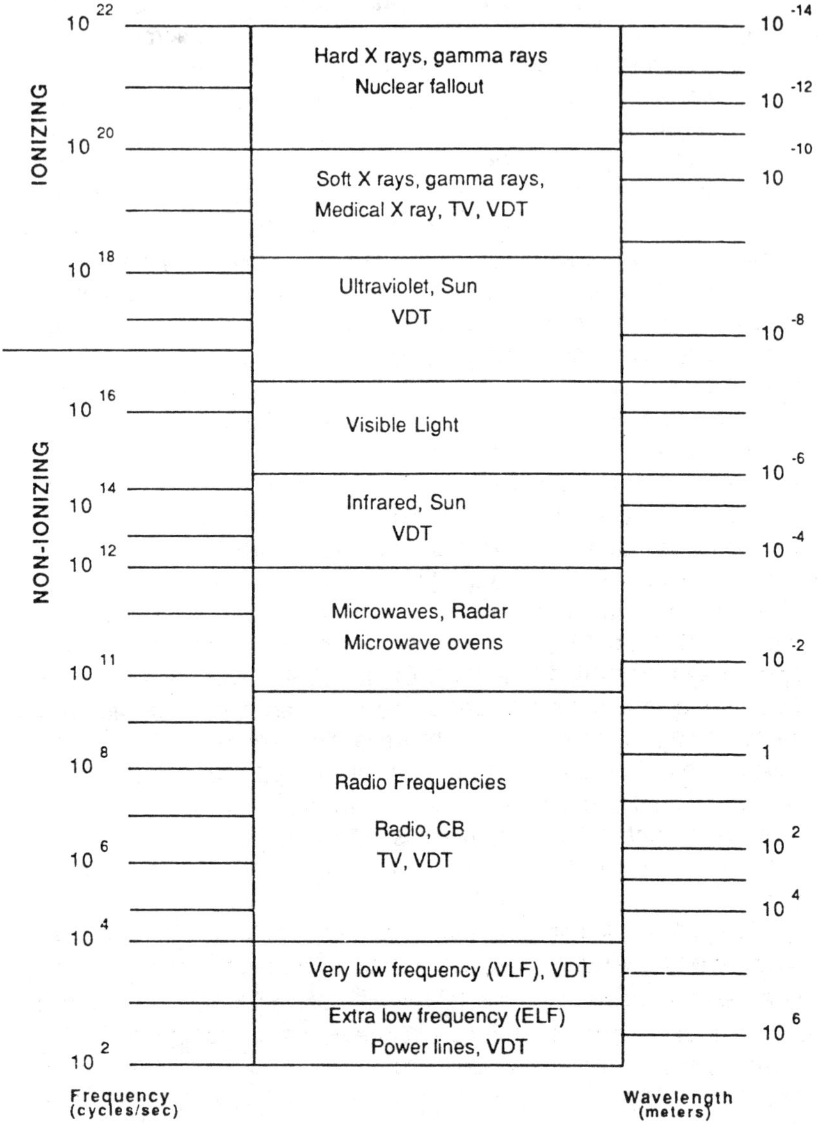

The Electromagnetic Spectrum

CHAPTER 2:

LOW-LEVEL IONIZING RADIATION

What is low level radiation and is there a safe dose which is considered harmless? Low level radiation means low doses of ionizing radiation. (See Chapter 3 for a discussion on low levels of non-ionizing radiation). In Dr. John Gofman's most recent book, ***Radiation-Induced Cancer from Low-Dose Exposure,*** he defines "low dose radiation" as internal-organ doses below about 20 rads, which he states is "*the low-dose range for most environmental, occupational, medical (diagnostic), and dental exposures.*" Until recently, the main stream media has emphasized only the effects of ionizing radiation, from nuclear bombs and radioactive fallout from nuclear plant accidents. We have been repeatedly told by the radiation industry that low levels of ionizing or non-ionizing radiation are harmless and may even be healthy for us. This is despite over thirty years of research by respected doctors, physicists and scientists to the contrary.

To understand low level radiation, imagine receiving a dose of radiation equivalent to an x-ray. This can vary from 10 to 1000 millirads, depending on where it is (dental, pelvic, chest, etc). Instead of absorbing the dose of x-ray in an instant, imagine spreading the dose out and receiving it in little increments over a long period of time. Or, imagine yourself living within thirty miles of a slightly leaky nuclear reactor. (They are all slightly leaky). This is what I mean by protracted low level ionizing radiation. Which is worse? According to Dr. Abram Petkau, a Canadian scientist, protracted low level ionizing radiation is worse. In other words, daily fallout is more harmful than getting the equivalent low dose in an x-ray.

Dr. Petkau studied the effects of very low level radiation - an innovative study since most people believed until recently that only high level radiation is harmful.[5] He found that it takes a mere 700 millirads of protracted radiation to break a cell membrane. Why does very protracted low dose radiation produce unexpectedly large free radical damage compared to short exposures of the same dose? Because the free radical process becomes extremely efficient at low levels of radiation. I want to emphasize Dr. Petkau's observation that **the more protracted the radiation dose, the lower the dose required to break the cell membrane.** This chilling observation explains why less people than anticipated died during the immediate time after the Chernobyl explosion and Hiroshima whereas leukemia and other cancers are occurring among children and adults at 100-1000 times the predicted rate based on the Hiroshima data. At high doses of radiation during a short time period, the free radical process becomes very inefficient due to the extremely large numbers of free radicals generated per unit volume. They are literally so reactive they smash into each other and wipe themselves out. Not to alleviate your worries, however, another, equally insidious direct action of radiation on the DNA in the cell dominates, producing mutations that lead to cancer, also a free radical process.

What is this free radical produced from low dose radiation and how is it produced? It is superoxide anion radical ($O_2^{\cdot -}$), one of those nasty free radicals we mentioned earlier. This free radical is produced from dissolved oxygen in water which captures an electron liberated by the absorbed radiation. Superoxide anion heads straight for the cell membrane and destroys it. Dr. Sternglass says "*since the chance of reaching the cell-membrane is greater when there are few other free radicals present, the efficiency of this process increases as the dose-rate declines.*" [6] Natural background radiation causes damage by the same mechanism. It is ironic that oxygen, a life giving substance is also the precursor of a deadly free radical that can destroy us cell by cell.

Oxygen-breathing organisms would never have survived if it weren't for the antioxidant dismutase enzymes, superoxide dismutase (SOD) and its partner, catalase. These enzymes convert superoxide anion first to hydrogen peroxide and then to water and oxygen. Lucky for us!! If it weren't for these enzymes, we would have never survived background radiation, let alone nuclear fallout! Read the supplement section to find out where to get these antioxidant enzymes.

Our daily dose...

To get a realistic view of radiation, let's compare natural background radiation with what we receive from man-made sources. Radiation is measured in REMS (roentgen equivalent man) which measures the degree of biological damage caused by ionizing radiation and takes into consideration the relative biological effectiveness of the absorbed energy in living tissue. It is also measured in RADS (radiation absorbed dose) which measures the radiation energy actually absorbed by the body. For all practical purposes, these are equivalent.

The latest estimated yearly radiation dose received by Americans has been increased from 170 to 360 millirems and for the first time includes radon, a natural radioactive gas from the radioactive decay of uranium. Normal background radiation includes approximately one third each natural radioisotopes, external man-made radiation and cosmic rays. A commercial pilot absorbs several hundred millirads flying coast to coast. Average radiation doses from medical x-rays include: 300-500 millirads for pelvic x-rays, 10-500 millirads for chest x-rays and 100-1000 millirads to the face for a full set of dental x-rays. The permissible maximum allowable radiation for the general public is 500 millirems (which includes background radiation plus 25 millirems from nuclear plants). But the Department of Energy (DOE) and the Nuclear Regulatory Commission (NRC) allow 5,000 millirems (about 5 rems or 5 rads) occupational exposure for nuclear industry workers. This is the common feature of all polluting industries - to increase the allowable exposure. Please do not confuse "safe dose" with "permissible dose". **The permissible dose is what you're going to get anyway and it has nothing to do with health or safety.**

Dr. Karl Z. Morgan, director of the Health Physics Division of Oak Ridge National Laboratory from 1943-1972, wrote in the *Bulletin of the Atomic Scientists* (September 1978), " *there is no safe level of exposure and there is no dose of radiation so low that the risk of a malignancy is zero...the genetic risks and especially those associated with recessive mutations may be as harmful and debilitating to the human race as the increase of cancer."*

There are many distinguished scientists, physicists, epidemiologists, doctors, and researchers who have studied the health effects of radiation since we started using it some 50 years ago. People like Drs. John Gofman, Ernest Sternglass, Alice Stewart, Abram Petkau, Karl Z. Morgan, Thomas Mancuso, Paul Brodeur and many others. For example Dr. Gofman says that 5-15 rems is the "doubling dose", the dose required to double the rate of cancer. Note that 5 rems, or 5,000 millirems is what nuclear plant workers are allowed. Others of the above scientists came to the same conclusions: that we have grossly underestimated the health effects of radiation, especially protracted low dose radiation; and that there is no such thing as a safe dose. But most of these people got what I call *three-D'd*: debunked, defunded and dismissed, because the truth about low level radiation is a great threat to the nuclear industry.

With permission from Sidney Harris.

"First, we have to convince the people that good health isn't everything."

CHAPTER 3: RADIOACTIVE BY-PRODUCTS FROM NUCLEAR FALLOUT

Contrary to what many people believe, radiation is far more toxic than any chemical, pesticide or other poisonous substance. Why? It's a matter of energy. Any form of energy sufficient to break the chemical bond can produced free radicals. But ionizing radiation is sufficient to rip electrons out of their orbits. According to Dr. Ernest Sternglass, each electron emitted by a radioactive nucleus has several million electron volts of energy, "*sufficient to disrupt millions of organic molecules in living cells. As a result, radioactive isotopes which concentrate in specific organs (such as iodine 131 in the thyroid), are much more damaging than ordinary chemical toxins such as lead or teratogens such as thalidomide.*" [6]

Secondly, radiation doesn't go away quickly. It is not biodegradable like your everyday garbage. Instead, each radioactive substance has its own half-life, the time it takes for one half of it to disintegrate via the emission of radioactive particles. This is can occur inside your body if you breathe radioactive air, eat radioactive food, or drink radioactive water. To give you an example of what I mean by half life, radioactive strontium (strontium 90) has a half life of 28 years. This does not mean it disappears in that time. To find the full life of a radioactive substance, multiply its half life by 20. So strontium 90 will be around for 28 x 20 = 560 years. Plutonium 239, with its half life of 24,400 years, will be around for 500,000 years. Not a pleasant thought when your house is built on top of radioactive landfill!

Radioactive isotopes are dangerous because they emit radioactive products as they "decay" to a more stable mass, and because certain isotopes tend to concentrate in certain organs. Inside the target organ, the isotope decays, causing cellular damage such as toxin production, enzyme (protein) inactivation or destruction, cell membrane damage (altered cell permeability), altered cellular metabolism, mutations, and abnormal cell division (inhibited, retarded or cancerous).

The isotope most feared is iodine 131 which tends to concentrate in the thyroid gland and the ovaries. It has an eight day half life which means that its radioactivity decreases by 50% every eight days. That's a rapid decay rate but it also means that the iodine releases heavy doses of radioactivity. Radioiodine is especially dangerous to the unborn and to young children whose growth depends on a healthy thyroid gland.

Also, children absorb other toxic metals, such as lead, five times faster than adults until they are past the age of five.[7] It is important to know that the absorption of toxic substances is inhibited in well nourished children and adults, and increased in malnourished people, including those who eat processed (junk) foods. Why? Because the same enzymes, vitamins, minerals, and amino acids that are busy forming your cells, running your body chemistry and producing energy are also involved in detoxifying the body and protecting it from the effects of radiation, toxic chemicals and pollution. For example, calcium inhibits the absorption of lead and prevents its deposition into bones and teeth. Radioactive iodine will not concentrate in a healthy thyroid containing the optimum amount of iodine. This is called the theory of selective absorption. It means the body first absorbs whatever nutrient minerals it needs and if they aren't there (due to eating junk foods), the body will selectively absorb similar substances, even if they're toxic, or radioactive counterparts.

How does radioactive iodine and other radioactive substances get into our bodies? From inhaling air, ingesting food or drinking water containing it. For example, milk-producing animals dine on grass laced with iodine 131 and other radioactive materials. It concentrates in them and then in the human thyroid as we drink the milk and eat the beef.

Two other important isotopes are cesium 137 and strontium 90. Both are abundant and have long half lives, 30 years for cesium 137 and 27.7 years for strontium 90. Cesium 137 is deposited in muscle tissue and in the ovaries whereas strontium 90 is deposited in bone. The targeted organ is irradiated for months or years, increasing the risk of cancer at that site. For example, strontium 90 increases the risk of bone cancer (multiple myeloma) and blood diseases, including leukemia. Major isotopes and their target organs are listed at the end of this chapter.

More details about this can be found in Dr. Steven Schechter's book, ***Fighting Radiation with Foods, Herbs and Vitamins,*** and Sarah Shannon's book, ***Diet for the Atomic Age***. Theodore Taylor, the physicist who helped investigate the Three Mile Island accident (1979) said that the amount of strontium 90 and cesium 137 emitted by the Chernobyl nuclear plant (in the worst case) could be up to 2,000 times larger than the volume released by the Hiroshima bomb and that a single reactor the size of the one at Chernobyl had enough radioactive material to cause more cancer deaths than Hiroshima. An estimated 70,000 Japanese people died within one month of the bombing of Hiroshima on August 6, 1945.[8]

IONIZING RADIATION

TARGETED ORGAN	ISOTOPE
Thyroid	Iodine 131
Skin	Sulfur 35
Lungs	Radon 222 Uranium 233 Plutonium 239
Liver	Cobalt 60
Kidneys	Ruthenium 106
Ovaries	Iodine 131 Cobalt 60 Ruthenium 106 Cesium 137
Bone	Radium 226 Strontium 90
Muscle	Cesium 137

How ionizing radiation is concentrated in the human body

CHAPTER 4: RADIATION AND THE UNBORN

Cancer is now the leading cause of childhood death until the age 14 and kills one person approximately every minute. How much of this do you think is caused by radiation of the young and the unborn?

A 1959 University of Michigan study by Beirwaltes showed that the small fetal thyroid concentrates iodine 131 ten times more than the infant thyroid and 100 times more than the adult thyroid.[9] This means that the internal dose to the thyroid gland of the fetus derived from the maternal diet and drinking water can be thousands of times greater than the external dose from ground radioactivity measured with a Geiger counter sensitive only to the penetrating gamma radiation.

Dr. Sternglass documents the dangers of radiation to the unborn in a review article on "The Implications of Chernobyl for Human Health".[6] The following information is taken from this excellent article. The highest amount of radiation reported by the EPA after the Chernobyl accident was in Vienna on April 30, 1986 where rainwater measured 1.7 million pCi per liter and milk reached 50,000 pCi per liter. (One pCi is one trillionth of a curie and one curie is an amount of radioactivity equal to one gram of radium or 37 billion disintegrations per second). If this is so, Sternglass states that *"the total dose to the thyroid of a fetus in Vienna following the accident could have been as large as 240,000 millirads if the mother had continued to consume the local milk, a dose comparable to those received by the children in the Marshall Islands who were accidentally exposed to the fallout of a large hydrogen bomb test on March 1, 1954 carried out at Bikini."* Sternglass further reports that several studies have shown that a dose of only 100 millirads to a fetus during the first trimester *"doubles the risk of developing leukemia or cancer. This is a dose equal to that delivered to the fetal thyroid by a mere 250 pCi of I-131 taken in by the mother, corresponding to the disintegration of only 550 atoms of iodine per minute."* Consider these realistic examples: A fetal thyroid dose as high as 20,000 millirads could have occurred from a pregnant woman drinking only one liter of milk following the Chernobyl accident. The average amount of milk that a one year old infant drinks per year could yield a dose of 1,500 millirads to the infant thyroid.

There are some graphs in Sternglass' review article which

strongly correlate cancer mortality, infant mortality, and increases in underweight births with nuclear fallout from large-scale nuclear weapons tests. Even subtle damage during fetal development can lead to immune suppressed children with retarded physical and mental capacity. Sternglass lists four references which report a decline in Scholastic Aptitude Test scores exactly 17-18 years after the onset of large-scale nuclear weapons tests, **in direct relation to the measured amounts of iodine-131 in the milk.**

What kind of legacy are we creating for our children, in terms of optimum health and mental development, when government's highest priority is a large nuclear industry with nuclear reactors (on land, in space and in the ocean), food irradiators in every state, continuous production of nuclear weapons and nuclear weapons tests around the globe?

Pregnant Women Exposed to Excess Fallout of Ionizing Radiation

Children born of mothers exposed to excess radiation have birth defects such as abnormally small heads, mental retardation (Down's Syndrome) and leukemia. A small head, observed in some Japanese children born in Hiroshima and Nagasaki after the bombs, indicates a failure of the brain to develop fully. Intelligence will be slightly or severely reduced depending on the severity of the brain depletion. A research team at Johns Hopkins University found that combined radiation exposure of mothers of children with Down's Syndrome in Baltimore was seven times greater than that of mothers of normal children, indicating the cumulative effect of radiation. They also found a higher rate of Down's Syndrome children whose fathers had been exposed to radar radiation. Another scientist found that many females whose mothers were exposed to radiation experienced a "superovulation". These women were often in such a continuous state of pregnancy that by the age of 28, 10% of them chose surgical sterilization. Many of these women also had general poor health as they approached their thirties.[10]

X-Rays and the Unborn

Gofman is especially concerned about children, who are far

more susceptible to radiation-induced cancer than adults. Dr. Gofman's conclusions are not based on new data. In 1958, Dr. Alice Stewart (Oxford University) published the results of large-scale epidemiological studies. She found that children who received only a few x-rays during fetal development had twice the risk of developing leukemia and other cancers before age ten than children who did not receive intrauterine x-rays. The average pelvic x-ray dose is 300-500 millirads. This means that doses as small as 1,000-1,500 millirads during the last months of pregnancy could double the rate of childhood cancers.[11,12]

What about fetuses irradiated during the first three months of pregnancy? By 1970, Dr. Stewart reported the results of a study of 16 million children born in Europe. In this study she discovered that fetuses irradiated during the first trimester had a cancer risk ten times greater than those irradiated just before birth. This means that the young fetus only needs 100 millirads to double his/her normal risk.[11,12]

Does Nursing Provide Protection From Low Level Radiation?

Nursing is especially important for babies exposed to low level radiation. (P.S. all babies are exposed to low level radiation). Breast-feeding counteracts radiation. Breast-fed infants excrete more strontium 90 than they ingest and more than bottle-fed infants.[13] This is because mother's milk has a lower phosphorus level than cow milk and phosphorus increases the retention of strontium 90. Thus, breast milk with its low phosphorus content allows baby to excrete strontium 90. The same is true for other toxins such as PCB's. Even though breast milk sometimes contains higher levels of PCB's than cow milk, PCB excretion is higher in breast-fed infants than bottle-fed infants.[14] See Chapter 20 for more information on nursing.

CHAPTER 5: X-RAYS

Anyone who wants to know more about the effects of radiation from dental and medical x-rays on human health should read Dr. John Gofman's book, ***Medical and Dental X-Rays: Health Effects of Common Examinations***,or a three-part article in *The Dental Assistant* entitled "Ionizing Radiation: Concepts for The Dental Assistant."[15]

In these articles he presents a strong case for reducing the radiation dose from dental and medical x-rays and for testing x-ray equipment in each office. Indeed, Dr. Gofman estimates that more than 45,000 fatal cancers are induced each year from x-rays. He further states that if the average x-ray dose were reduced three-fold, about one million cases of cancer per generation could be averted. Betsy Russell-Manning discusses this topic in both of her books on silver amalgams and candida albicans (A).

Let me paraphrase some statements in an article by Paul Brodeur in *US Magazine,* December, 1986.[4]

* Children x-rayed in-utero have close to 50% greater risk of developing cancer before age 16.

* Most pulmonary physicians advise against annual chest x-rays except in heavy smokers or other high risk groups such as asbestos workers.

* The dose with mammograms has been sharply reduced and they are given more selectively now because it is now realized that the radiation delivered during mammography might cause more cancer than the exams detected. Also, there are false negatives.

* Regarding dental x-rays: they are relatively safer because they have a lower degree of penetrability. Some experts believe that an annual dental x-ray is unwise, especially in children. The American Dental Association has advised against routine full mouth dental x-rays.

Did you know that the older 50-kilovolt x-ray machines produce clearer x-rays than the 90-kilovolt x-ray machines currently being sold by the industry. Perhaps public awareness of and concern radiation dangers will encourage the industry to manufacture the lower voltage machines.

CHAPTER 6: THE MYTH OF THE SAFE NUCLEAR PLANT

We have a lot to learn in the safety department, as some study and research groups such as the Fund for Constitutional Government and the Critical Mass Energy Project (a Ralph Nader Study Group) have concluded. Two recent studies, one by the Nader group and one by Rep. Edward Markey (D-Massachusetts) concluded that U.S. nuclear safety is worsening in regard to the number and severity of mishaps. Under a new reporting system devised by the nuclear Regulatory Committee (Which yields about half as many accident reports as the old one, says the Critical Mass report), there were 2,417 reports on mishaps filed in 1984 and 2,974 mishaps filed in 1985. This increase parallels the increase in operating reactors, from 68 in 1979 to 100 in 1986. [17]

A 1983 study by the Fund for Constitutional Government concluded that the Navy has systematically suppressed information on nuclear accidents including spills of radioactive materials and other environmental and health problems.[16,17]

On July 12, 1990, in the *San Jose Mercury* and other newspapers across the country, DOE Secretary, James Watkins, said that a study financed by his agency admitted huge radiation releases in the 1940s and '50s. For example, in 1945, infants living a few miles downwind from the Hanford nuclear reactor, could have received radiation doses as high as 2,500 rads to their thyroids.

The History of Rancho Seco

To illustrate the tremendous problems of the nuclear industry, let me review the history of Rancho Seco Plant south of Sacramento. I chose Rancho Seco because it's the first operating nuclear power plant to be voted out of business by its customers, those of the Sacramento Municipal Utility District (SMUD), in June 1989. In more than a dozen similar referendums elsewhere in the nation, voters chose to keep plants operating. Rancho Seco's fifteen year history was clouded by nearly one hundred unplanned or accident-forced shut-downs during the last eleven years, a number twice the national average. In fact, since 1975, Rancho Seco has been shut down 60% of the time. The Nuclear Regulatory Commission (NRC) rated Rancho Seco the sixth-worst run nuclear power plant of 101 operable commercial nuclear reactors in the United States. The other five had already been shut down. An NRC investigation found that SMUD had virtually abandoned preventative maintenance from start up until 1986. This is similar to buying a new car and having no tune-ups, oil changes or lube jobs for the first 150,000 miles. Equipment problems and operator errors led to some of the worst over-coolings on record.

On December 26th, 1985, the Ranch suffered its most serious accident, which led to its final closure in 1989. A relay circuit in the computer that controlled operations of the plant shorted out, causing complete loss of control of the reactor for nearly 30 minutes. Workers narrowly averted a meltdown by manually controlling valves in the system. The worker, who successfully figured out the sequence that saved the reactor, collapsed and had a mental breakdown following his experience. His burden was too much considering the plant did not have the phones of back-up staff on site, was understaffed and unable to carry out prescribed emergency procedures including communications with outside emergency officials.[18]

In NRC's follow-up investigation, the worse discovery was not just the lax training programs for the staff but the substandard state of maintenance. One major valve involved in the 1985 accident had not been serviced for eleven years. The seriousness of this accident did not surface until after the Chernobyl accident four months later (June 1986).[19]

Other news surfaced in 1989 just days before the vote to close the plant: in 1987 the NRC found the Rancho Seco plant operators doctored documents to conceal the release of radioactive water from the facility into nearby Clay Creek. NRC also found that SMUD had altered results of radioactivity tests to make them acceptable and had redesigned the plant without NRC approval.[20]

This story is not unique or unusual. Only the name of the reactor, its location and the nature of the problem varies. But whether the reactors leak, break down, melt down, or explode, the effect is always the same: premature death and decreased quality of life due to radiation-induced cancer and other diseases resulting from immune system damage.

Three Mile Island - America's Chernobyl

TMI started operating in 1974. The accident occurred on March 28, 1979 in the early morning hours. It happened because of human error, valve malfunctions and a close call with meltdown in Unit Two that sent radioactive emissions into the air and water of central Pennsylvania. On the same day, Dr. Ernest Sternglass initiated a debate over precisely how much radiation was unleashed from the accident by arriving at the Harrisburg airport with a Geiger counter and charging that the levels were much higher than the Metropolitan Edison power company had claimed. Actually, the highest estimates for the initial surge of radioactive emissions immediately following the accident came from Met Ed itself, whose operators reported radiation as high as 40 rads per hour falling on nearby Goldsboro. Met Ed now claims that their emission records during the first two days of the accident have been lost or destroyed.

Some time later in March, 1979, Mrs. Vera Chatz, a resident near Three Mile Island called the Harrisburg Patriot and learned that 20 millirems per hour of radiation had been detected at Goldsboro, Pennsylvania, a town opposite the island on the river where the nuclear plant stands, about one mile away. A chest x-ray delivers 20-500 millirems, yet the people of Goldsboro were being exposed to 480 millirems every 24 hours.

The Nuclear Regulatory Commission was not told about the Three Mile Island accident in 1979 until 56 hours had passed. Public relation statements about TMI hastened to point out that no adverse health effects had been observed. Met Ed claimed that the accident had delivered a harmless X-ray-equivalent dose to the entire area, including the pregnant women. Yet the residents of central Pennsylvania experienced all of the symptoms of high exposure radiation sickness (metallic taste, hair loss, open sores, high white blood count, etc.). By the way, to get the metallic taste, millions of picocuries of iodine had to be present. To get skin reddening, rashes and hair loss requires a 300 rem dose.

Now that the latency period for developing cancer and other diseases following exposure to radiation has passed, much data has been collected by concerned epidemiologists and a lot of this has been reported by Harvey Wasserman who co-authored ***Killing Our Own: the Disaster of America's Experience***

with Atomic Radiation. Here I will just summarize some of the health effects reported. The cancer rate increased by seven times. From 1979 to 1984, 11 out of 19 births (58%) were faulty. The list includes spontaneous abortion, pregnancies outside the uterus, low birth weight babies, still births, and an epidemic of neonatal hypothyroidism.

Other health effects include an increase in leukemia, large tumors, rare muscle tumors (cesium 137 goes into muscle tissue), allergies, multiple sclerosis, asthma, bone marrow suppression, increased white blood counts, cataracts, immune system disorders, rashes, sudden pain causing the victim to double over for no apparent reason, unexplained fatigue, bizarre sores that won't heal, etc. Plants and animals suffered the effects of radiation too. For example, entire gardens turned black, an ancient oak tree wilted overnight, and strange plant mutations occurred, such as dandelions with three foot leaves, 55 pound mushrooms, cantaloupe-sized tomatoes. Farm animals suffered decreased fertility, increased spontaneous abortions and increased cesareans from 1-2 per year to 1-2 per week.

Newborn calves could not stand up. Wild birds died by the dozens. Litters of cats were born deformed. Groups of animals died suddenly en masse shortly after the accident, such as more than a dozen cows on one farm and 500 exotic birds at an aviary.

To date Met Ed has paid an estimated $20 million in personal injury claims to the people of central Pennsylvania, 2,000 of whom have sued for damages. Met Ed made silence a prerequisite for about 200 cases settled out of court so only cases involving children or estates of the deceased requiring hearing by a judge have become public knowledge. Met Ed's credibility was brought to question when it admitted to one felony count of falsifying documents related to discharging radioactive water prior to the accident and "no contest" to six additional charges.[21]

Studies on the Effects of Radiation

In 1962, the Atomic Energy Commission hired Dr. John Gofman to investigate the impact of radiation on human beings at the Lawrence Livermore Laboratory in Northern California. In 1969, Dr. Gofman and Arthur Tamplin concluded that radiation produces a linear increase in cancer incidence, and children are far more susceptible to radiation-induced cancer than adults. They concluded that the official U.S. permissible average dose of radiation from nuclear plants (at that time 170 millirems per year) would result in an additional 16,000-32,000 cancer deaths annually. The AEC defunded Gofman and Tamplin's project in 1972 and have defunded other scientists who have predicted or shown cancers or genetic defects from "permissible" limits of radiation exposure.

The Department of Energy (DOE) replaced the AEC in 1975 and has continued the AEC tradition of discrediting and defunding medical research scientists whose conclusions conflict with its nuclear energy programs. For example, DOE defunded Dr. Rosalie Betrell's Tri-State Leukemia Survey which linked x-rays to leukemia. DOE attacked Dr. Ernest Sternglass' professional competence when his research (funded by the University of Pittsburgh Medical School) correlated increased infant mortality with nuclear plant radioactive releases.[16]

DOE then hired Dr. Thomas Mancuso to investigate a report by Dr. Samuel Milham (Washington State Department of Health) that there were elevated cancer rates among Hanford, Washington nuclear reactor workers. Dr. Mancuso reported a 6% increase in the cancer among the workers. DOE defunded Dr. Mancuso.[22]

Uranium mines, the source of nuclear fuel, generate radon gas which is inhaled by miners and the surrounding townfolk and animals. A U.S. Public Health Service study reported a fivefold increase in cancer deaths among miners compared to the unexposed population. In March 1979, a truck spilled 7-1/2 tons of processed uranium ore on its way from Casper, Wyoming to the Kerr-McGee enrichment plant in Gore, Oklahoma. Do you think this is an uncommon incident? [16]

Frank Shiavo, a lecturer in environmental studies at San Jose State University said *"nuclear power produces only 13% of our nation's electricity, which amounts to 3.5% of our total energy. It is one of the most costly ways to generate electricity--at 10-12 cents per kilowatt hour."*[16] This means that nuclear reactors cost twice as much as other sources of energy.

At a news conference in Santa Cruz, California on Thursday, May 15, 1986, Commissioner James T. Asselstine, one of the Nuclear Regulatory Commission's five commissioners said that a recent NRC study concluded that there is a 45% chance of a severe core meltdown accident at one of the United States' 100 operating nuclear plants during the next 20 years.

Chernobyl's Legacy

Using computer projections, scientists are predicting a grim legacy for as many as one million Russians, Europeans and Americans who may develop cancer in the next several decades. These predictions are based on estimates of the amount of radiation resulting from the accident and estimates of greater than previously assumed risks from low level radiation - especially protracted low dose radiation (e.g. received over an extended time period).

According to a study by Lawrence Livermore National Laboratory in Livermore, California, the Chernobyl nuclear *accident released "as much long-term radiation into the world's air, topsoil and water as all the nuclear tests and bombs ever exploded."*[23] Further, the report says that this long-term radiation may contain 50% more cesium 137 than the total radiation produced by all atmospheric tests. Cesium 137 does not decay into harmless products for more than 100 years - a stiff price to pay for a nuclear boo-boo.

Gofman, estimated that one million people including over one half million outside the Soviet Union will develop cancer as a result of the Chernobyl accident. Half of these cancers will be fatal. Dr. Gofman's estimates project the number of deaths per country as well.

Dr. Sternglass' estimates for Europe are similar: up to 600,000 additional cancer cases resulting from Chernobyl.[24] Other scientists have predicted much lower casualties because they have not considered Petkau's research on the damaging effects of very low level radiation discussed in Chapter 2.

The Committee for Nuclear Responsibility highlights the following statements. The radiation industry told us that the cancer deaths from Chernobyl will be undetectable. This is true because the cancer rate is so high that 500,000 cancer deaths are within the statistical error belt. This means that one out of three people get cancer and one out of six die from it. This means one death every minute. **Is it okay to kill one half million people as long as your victims remain undetectable? Is undetectable the same as insignificant?**

The radiation industry says that many more people will die from spontaneous cancer than from Chernobyl. True again. Spontaneous cancer kills one out of six people. Nothing competes with it as a killer. Even war is a lesser killer than the "natural" cancer rate. **Is it socially acceptable to kill people if you do it at a slower rate than the spontaneous rate of cancer deaths?** [25]

What will the United States inherit from Chernobyl? Nick Heffter, a research meterologist with the National Oceanic and Atmospheric Administration said in a telephone interview from Washington that fallout from the Chernobyl accident could cover the U.S. for years. Let's look at two water samples taken Monday, May 5, 1986, one in Portland, Oregon and one in Richland, Washington. The Portland sample showed 630 picocuries of radioiodine 131 per liter of rain water. The Richland sample registered 500 pCi per liter. One liter is about six wine glasses of water. Drinking one liter of the Richland rain water would be equal to a little less than one-half a chest x-ray. This means that you get a chest x-ray dosage of radiation every time you drink 12 glasses of Richland rain water. Would you consider this safe?

By the way, EPA's water standard permits only three pCi per liter for consumption on a long term basis but you can give your kids 15,000 pCi per liter of milk if you only do it once a year! [27]

The Future

Do not think we are safe from new nuclear plant start-ups. In a San Jose Mercury newspaper article by Mike Langberg entitled, "Nuclear Industry Tries for Comeback in 1990's", I read that *"the nuclear power industry is quietly planning a comeback in the 1990's through a powerful coalition of utilities, government regulators and engineering firms."* [28] In September 1989, this new coalition received $100 million from the U.S. Department of Energy to finance the development of a new generation of nuclear reactors, some of which could be on stream before 1990. Known by the acronym APLWR (advanced passive light-water reactor), the new reactors would be about one half the size of first generation nuclear power plants and would, according to the industry, be much safer and simpler to operate. Yet, "these are the same promises the nuclear industry made 20-30 years ago that were completely untrue," says Daniel Becker of the Sierra Club. In fact, the new designs are still vulnerable to meltdowns, defective construction, corrosion of steam pipes, poorly trained operators and the same waste disposal problems. [28]

CHAPTER 7: IRRADIATED FOOD OR, WHAT TO DO WITH ALL THAT NUCLEAR WASTE!

What is Food Irradiation? A technology designed to use radioactive waste products from the nuclear industry to irradiate meats, grains, fruits, vegetables, herbs and spices.

The Plan: To build 2,000 food irradiation plants by the year 2000 which will irradiate food with radioactive cobalt 60 or cesium 137 or linear accelerators using massive doses - between 100,000 and 3-million rads, depending on the food. By comparison, a chest x-ray gives off less than one rad (from 300-1000 millirads).

The Goal: To irradiate almost **half** the food Americans eat by the end of this decade.

The Stated Reason: To keep your food fresh (?) and free from bugs (fungi and bacteria). We are told that food irradiation is necessary to preserve foods for the Third World countries and to protect food from toxins. However, **aflatoxin** (a highly carcinogenic substance from food molds) is produced in greater quantities in irradiated food and it is this toxin that wipes out most of the Third World stored foods. **Botulism,** a deadly source of fatal food poisoning, is not killed by radiation, but its natural enemies are. Now, thanks to food irradiation, food may be contaminated without any warning smell.

The Real Reason: The U.S. Government and the Department of Energy are faced with enormous volumes of radioactive waste, which are exceedingly costly to store. The State of New York, to give an example, spends three million dollars per year to maintain these wastes.

Under growing pressure because of the increasing amounts of nuclear waste and no program for disposal or permanent storage, the DOE has become our largest subsidizer of food irradiation. For example, in 1985, congress authorized DOE to spend $15.6 million dollars to support food irradiation.

On Independence Day, 1990, I read in the *San Jose Mercury News,* the latest DOE estimate for nuclear cleanup costs to be **$31 billion** over the next five years. This figure represents a 50% increase over 1989 projections.

The Hazards: This is not new information. The following excerpt from research has been known for at least 20 years. Gamma radiation causes the formation of little understood chemical compounds in food which have been named **unique radiolytic products (URPS)**. URP's have not been fully studied and the work so far done is not reassuring. Canada's Pollution Probe Foundation cites animal studies showing that some animals fed irradiated food developed extra sets of chromosomes. A March 1984 USDA report on 12 studies showed that animals fed irradiated food developed increases in testicular tumors, lesions such as cancer and kidney disease, and reduced life spans. Gamma radiation depletes or destroys important nutrients in foods including vitamins A, E, K, B-1, B-2, B-3, B-6, B-12, vitamin C and essential amino acids cysteine, methionine, histadine, and tryptophan, as well as enzymes. There is not a single component of food - protein, amino acids, fats, carbohydrates and enzymes - that remains chemically unchanged when irradiated. Some of these food irradiation byproducts are toxic and/or carcinogenic and mutagenic.

Most of the objectives to food irradiation are insignificant when compared to the **radiomimetic** factors. The process of irradiating food forms radiomimetic chemicals - which mimic or act like radioactive substances. Radiomimetic chemicals can injure the chromosomes either by altering the chemistry of a single gene so that the gene conveys improper information (point mutation) or by actually breaking the chromosome (depletion). Some types of cell damage may lead to uncontrolled cellular growth which in humans is called cancer. Phenoxy herbicides and other "farm chemicals" cause cancer in the same fashion.

Radiomimetic chemicals may be responsible for some of the damaging effects observed in humans and animals who eat irradiated foods containing these chemicals. These effects include altered cell division (increased degeneration of cell nuclei, increased mitotic disturbances, increased chromosome breaks), mutations, decreased fertility in animals fed irradiated food (due to depleted enzymes, etc.), reduction in growth rate of offspring of animals fed irradiated food, reduced resistance to infections, increased spleen weight, and a significant increase in cancer. The effects on living tissue of eating irradiated foods are similar to those of direct radiation, even though irradiated food is not radioactive.[29-33]

The only study of humans consuming irradiated food published in the last 20 years is one done in India with starving children. The researchers abruptly stopped the study after only two months because 80% of the children in the irradiated-wheat group showed signs of chromosome damage (polyploidy) which was not observed in the group of children eating non-irradiated wheat. The researchers felt it unwise to continue the study since leukemia is a likely result of polyploidy.[34] A New Jersey parent, Bobby Miller, responded to a Radiation Technology, Inc. representative asking why he was on the anti-food irradiation side, "*If I follow their (anti-food irradiation) path and they are wrong, my children won't get hurt. If I follow your path and you are wrong, my children might die*".

Even if food irradiation were safe, perhaps the most serious drawback to the process is the increased use, handling, and transportation of radioactive materials to the plants. Food processors predict more than 1000 food irradiation plants in the U.S. by the mid 1990's. The cesium 137 will be stored in large pools of water or in the air when not in use. Cesium 137 is water soluble and a leak could contaminate our water supply. Radioactive water, like pesticides and other toxic chemicals, will seep readily into our ground water.

The Secret: The DOE, the Pentagon, and the U.S. Army refuse to release their research into the long-term effects of food irradiation, since it is considered "classified information in the interest of national security." Whose interest? Whose security? However, there is unclassified evidence that irradiated foods will drive food prices up 2-24 cents per pound, depending on the food.

Why aren't the food industry and nuclear waste disposal experts worried about whether Americans will hesitate to consume irradiated food? It's simple. **They are not planning on telling us what is irradiated and what is not irradiated**. Consumers are already eating unlabeled irradiated foods without knowing it. In April 1986, the FDA approved radiation doses of up to three million rads to control bacteria in spices and seasonings. These foods are sold nationwide without labels. No penalties will be administered by the FDA for failing to comply with the labeling requirements, nor does the FDA have a list of irradiators or irradiated foods.

The latest information on the label battle front as of this writing is that only whole foods will be labeled with a flower symbol called the "**radura**" and the words "treated with radiation". Any irradiated food combined with any non-irradiated food (such as water, sugar or salt) will not require a label. So if you buy a can of irradiated potatoes that contain sugar, water or salt, it will not be labeled. This means that about 80% of irradiated foods will never be labeled.

The Radura

The Three Mile Island of Irradiation - Decatur, Georgia, 1989: Radiation Sterilizers, Inc. is one of forty or so U.S. irradiation facilities which irradiate food, cosmetics, medical equipment and personal care products. In 1988, RSI, located in a densely populated area, had an accident which developed in a manner the experts declared impossible, has not been completely repaired as of July 1990 and has cost several million taxpayer dollars so far. Let me briefly describe this accident.

On June 6, 1988, a radiation leak was detected which automatically shut down the Decatur plant. It seems that one or more of the 252 capsules containing water soluble radioactive cesium had sprung a leak. These capsules were submerged in a 24-foot deep pool inside the RSI plant. More disturbing than "whose got the leak" was the admission by the State Environmental Protection Division official in charge of monitoring radiation hazards that he did not know about the plant's existence, twenty miles from the center of Atlanta. This is a curious admission, considering that RSI had been operating for three years and had the largest inventory of radioactive materials in Georgia other than the state's two nuclear power plants.

Back to "whose got the leak." The RSI executives found themselves with several problems, the first of which was: which of the 252 capsules of highly soluble radioactive cesium were leaking. They did not know. More than six months after the leak was first reported, two capsules were inadvertently discovered as the culprits, not because of any advanced technology but because they bulged suspiciously. They were removed and placed on a truck bound for Oak Ridge, Tennessee. This incident cost Georgia taxpayers five to seven million dollars. Naturally, Georgia state officials wanted the remaining cesium capsules to be removed from RSI premises for fear of further leakage. As of mid-July 1989, the stalemate continued with the RSI plant closed and 247 radioactive cesium capsules still in its holding pool.[35] About 100 capsules remain today. About 18 of these capsules were shipped to France in exchange for information on "Cesium into Ceramic Form" technology.

By the way, radioactive water contaminated about ten workers, who then contaminated the RSI office, their cars and their homes. Thousands of milk cartons and contact lens solutions made their way into the stream of commerce and were never recalled.

Current State of Food Irradiator Art

As of October, 1989 there are thirty-seven irradiators in operation, and eight which are not yet open. Five new facilities are currently under construction in El Paso, Texas, near the Mexican border, making it a convenient processing point for inbound food supplies from Mexico. Should food irradiation become widespread, it is believed that these border facilities could irradiate Mexican fruits and vegetables on their way to U.S. markets. Over 40% of the fruits and vegetables in the U.S. come from Mexico during the winter months.

Update on the Irradiation Eradicators

If you are concerned about this issue, there are things you can do. You can join the the Coalition to Stop Food Irradiation(A). Do not think that your efforts will be fruitless. Here are some grassroots success stories. A 1987 effort by private groups opposed to irradiation produced a 30,000-signature petition in Trenton, New Jersey which led to legislation to ban the use of food irradiation in the state of New Jersey.[36] This legislation passed the state Senate. Instrumental in this effort was a group called People for Responsible Management of Radioactive wastes and Food and Water, Inc. (A). As of this writing other states in North America have banned food irradiation, including Maine and New York.

On Thursday, December 6, 1988 the government of New Zealand banned food irradiation following an intensive review of food irradiation when an Australian firm submitted a proposal to build a gamma irradiation plant in Mangere, Aukland. The proposal was defeated by the citizens of Mangere armed with information provided by the National Coalition to Stop Food Irradiation.

Also the European Parliament voted on October 11, 1989 to ban irradiation of food throughout the European Community from the end of 1992. It voted to permit one exception to this general ban - for dried herbs and spices.

You can read more details on this hot topic in Tony Webb's book, ***Food Irradiation, Who Wants It?***

FOOD IRRADIATION HISTORY

Recent Major Events as of Summer, 1990

* In Washington, Alaska and Hawaii: funding has been cut off for previously planned irradiators.

* Oklahoma dropped plans for irradiation of food.

* Vindicators of Florida lost both potential irradiator plant sites because of public protest.

* The Florida Citrus Commission and the Washington State Apple Commission have rejected irradiation because of public pressure.

* Major corporations such as Beatrice/Hunt Wesson, Heinz, Quaker Oats, Amway, and Zacky Farms have declared policies of not using irradiation.

* Irradiation has been banned in New Zealand and in the European Economic Community.

* The states of Maine, New York and New Jersey have passed laws banning irradiation.

CHAPTER 8: NON-IONIZING RADIATION

A more common, and in fact, daily increasing form of electromagnetic energy is non-ionizing radiation (**NIR**). **NIR** is produced by just about everything in our high-tech modern life: electric current, radiowaves, microwave ovens, radar stations, television (cathode ray tube), video display terminals (VDT's), high voltage-lines, infrared and fluorescent lights, and sunlamps (e.g. tanning booths which emit ultraviolet light). VDT's have some of each type of radiation (ionizing and non-ionizing) but they are considered mainly NIR. I will discuss VDT radiation in Chapter 9 because it is so extensive.

Update On Low-level Non-ionizing Radiation

None of this information has been available from the main stream media until recently. There was a land-mark publication which came out in June, 1989 in three issues of the *New Yorker Magazine*, June 12, 19 and 26 written by Paul Brodeur. These articles deal with power lines, radar and video display terminals, respectively.[37] They are part of Paul Brodeur's book, ***Currents of Death.***

Power Lines

Electric Smog is a phrase used to denote non-ionizing radiation emitted by power lines, radio and microwave transmitters, and home appliances. The radiation emitted by most power lines, radio waves and microwaves has been thought harmless until recently as the results of research efforts have been collected. Consider these facts. The new ultra high voltage power lines have enough electricity in the air 200 feet from the wires to light a fluorescent bulb carried in the hand of a human being who stands near the high voltage source. Do you think it would be safe to live or work within a few hundred feet of these high voltage sources? The Russians don't. They will not allow buildings anywhere near these high voltage sources and they have strict rules over the amount and duration of microwave emissions from radio transmitters and radars a person can safely absorb. Not so with the U.S. We have no legal restriction for electromagnetic radiation exposure, only an informal guideline.

Wertheimer's Studies: Magnetic Fields from High Current Power Lines

In 1974 Nancy Wertheimer, an epidemiologist, began a study on the effects of power lines on childhood leukemia in the Denver area. This study was suppressed from the main stream media until 1989, when the results of her study appeared in the June 12th issue of the *New Yorker Magazine,*[37]

In the power industry, voltages as high as 230,000 are stepped down to 13,000 volts by large transformers at substations, producing high currents for distribution along primary wires to various parts of the city. Pole mounted transformers then step this voltage down to the 240 and 120 volt levels required to operate household equipment. Until the first service drop, these secondary wires produce relatively strong 60 Hz alternating magnetic fields up to a distance of 50 feet. High current primary wires produce strong magnetic fields up to a distance of 130 feet.

In her Denver study, Wertheimer found a two to three fold increase in leukemia rates for children who lived in homes near high current lines. One out of four homes were categorized as high current homes.

Even though magnetic fields are usually cancelled in ordinary wiring where the return current tends to balance the supply current, such cancellation is imperfect near dwellings because the wires are often separated in space and because some of the return current does not flow through the wires but is grounded through the plumbing system. Ground current flows not only in street plumbing but also through the pipes of houses and thus produces magnetic fields in these houses which is related to the wiring configuration near each house. Ambient fields in a house due to nearby distribution wires or plumbing may sometimes reach 2-3 milligauss levels for hours or days at a time.

Other data correlated with these findings. For example, A 1950 U.S. Public Health Service report correlated increased cancer deaths in workers who were frequently exposed to alternating current magnetic fields - power station operators, telephone linemen, power linemen, subway and elevated railway motormen, electricians and welders.

Wertheimer and Leeper theorized that alternating current magnetic fields might indirectly affect cancer development by depressing immune system response. The immune system might utilize electric potentials occurring at cell surfaces. When a human body is subjected to an electromagnetic field different from the level existing during man's evolution, the immune system may be altered.

High Voltage Power Lines and Pulsed Electric Fields

Dr. Robert O. Becker wrote about his studies of regenerative of bone healing in ***The Body Electric.*** He demonstrated that the regenerative process was initiated by an electric current associated with the nervous system. He discovered that bone fractures in humans heal in the same electrically controlled manner that governs the regeneration of limbs in a salamander (animals whose regenerative powers include the ability to grow new limbs). Using asymmetrical electromagnetic fields he stimulated the healing of fractures. From this work, he reasoned that if electric current could govern benign growth it might also play a role in the development of malignant growth.

Other researchers proved this. Andrew Marino, a biophysicist at the Veterans Administration Medical Center in Syracuse, New York, studied people and animals exposed to electromagnetic radiation simulating that around high voltage power lines. He reports, *"exposure levels like those under the wire can cause a stunting of growth. Levels like those 300-500 feet away cause physiological effects, such as changes in blood chemistry and heart rate. At 1000 feet, there are behavioral effects such as drops in human reaction time."* [3,37]

In 1973, W. Ross Adey of UCLA's Brain Research Institute, exposed laboratory monkeys to electric radiation of frequencies normally present around us every day. The monkeys exhibited behavioral changes and a distorted sense of time. Adey believes that electromagnetic radiation such as that emitted from high voltage sources creates artificial electromagnetic force fields that override the body's natural biological rhythms, the internal clock that regulates waking, sleeping and thousands of more subtle body processes. We respond to this artificial force field by adjusting our biological rhythms to the pulse of the electric smog. The resulting stress on the body decreases the body's general resistance (the immune system) and the person becomes vulnerable to diseases he might otherwise have resisted. This may explain why University of Colorado medical researchers found that the death rate for certain cancers such as leukemia is twice the average in homes within 130 feet of high-voltage lines.

Hundreds of Russian and Eastern European experiments have shown that electromagnetic radiation may be associated with many health problems including high blood pressure, heart disease, headaches, sexual dysfunction, and blood disorders such as a 50% increase in the white blood count.

How Does Low Level Non-Ionizing Electromagnetic Cause or Promote Cancer?

In the *New Yorker Magazine* article on radar (June 19, 1989) Paul Brodeur reported some interesting research that may help to explain this. There is an enzyme called ornithine decarboxylase (ODC for short), which is essential for growth in all cells. ODC participates in the synthesis of DNA protein. It has been known for some time that carcinogens increase ODC activity although not all substances that increase ODC activity are cancer promoters. Nevertheless, in clinical studies of cancer cells, high levels of ODC are a reliable index of malignancy.

In 1985, Ross Adey, Craig Byus and their colleagues discovered that low-level, pulsed, ELF-modulated radiation very similar to a newly developed type of radar surveillance system could increase ODC activity up to 50% in cultured human and animal cancer cells. This type of radar was installed in 1976 on Upper Cape Cod at the Massachusetts Military Reservation and in 1977 at Beale Air Force Base, 35 miles north of Sacramento, California. Here I will give only an overview of the research that demonstrates the dangerous nature of this type of pulsed low-level microwave radiation.

This new radar (called PAVE PAWS for Precision Acquisition of Vehicle Entry Phased Array Warning System) operated at a carrier frequency of between 420 and 450 MHz, with a pulse-repetition rate of 18.5 hertz and was very similar to the experimental conditions of the Adey-Byus research (450 MHz, pulse rate 16 hertz, power intensity of one milliwatt per square centimeter). Their extraordinary discovery was published in *Cancer Research,* August 1988. Lack of funding prevented more extensive experiments unless, paraphrasing Paul Brodeur, one believes that an experiment is being conducted upon thousands of human beings in the Cape Cod area surrounding PAVE PAWS who have been developing cancer at 17 times the normal rate for nearly a decade following the installation of the PAVE PAWS radar system.

Adey and Byus also found that when the ELF-modulated radiation was combined with a chemical carcinogen, their joint effect on ODC activity was double that of either alone. This demonstrates how interacting variables can produce wide variations in experimental results and why it's so hard to isolate the effects of each variable.

Dr. Adey and his colleague, Dr. Suzanne Bawin, a neurophysiologist, had discovered other disturbing effects of this same type of radiation long before PAVE PAWS was installed in 1976. They found that low-intensity microwave fields of 450 MHz modulated at the ELF frequency of 16 hertz altered the outflow of calcium ions in the brain tissues of living cats. This illustrates a direct effect of microwave radiation on the blood brain barrier and brain chemistry. Adey proposed a fascinating domino-effect model to explain how the tiny stimuli of weak electromagnetic fields might trigger amplified changes in brain chemistry. The cell membrane reacts in domino fashion to the arrival of a weak field by releasing calcium ions normally bound to the protein strands that protrude from the cell membrane surface. Because of the domino effect, the released energy is far greater than the initiating stimulus (low level radiation). This released energy is transmitted along the protein strands to the interior of the cell, where it stimulates or inhibits intracellular enzymes such as ODC and others. We will talk about the importance of enzymes in Chapter 14. Here I will only mention that enzymes are the workers of the body and are involved in all biochemical reactions in the living organism.[37]

This and other research clearly lays to rest the obsolete notion that the only harmful effect of microwave radiation is the heat it could create in the body if generated at high power, such as microwave ovens. *"Skeptics... called this obsolete exposure guideline, 'the English muffin standard,' because of the implication that microwave radiation was harmless as long as you didn't turn brown or feel toasty,"* (from Brodeur's ***Currents of Death***, page 130).

Microwave Ovens

The energy produced by microwaves is absorbed by molecules and creates heat in the material being radiated. That is why microwaves cook food. The heat created by the microwaves or radio waves in turn create molecular motion in the molecules, resulting in heat stress. Inside the human body, the same heating process occurs. The eye lenses are especially vulnerable because they lack blood vessels to help carry away the heat. Unusual cataracts have been found in the eyes of radar workers who are exposed to high levels of NIR. Unlike cataracts caused by ordinary heat, which form on the front surface of the lens (such as a dog gets after years of basking beside the fireplace), these cataracts form on the back surface of the lens, which is where radiation-caused cataracts would be expected to occur. Other organs that are particularly sensitive to microwave radiation are the gallbladder, digestive tract, urinary bladder and the testes.[38]

There is evidence from European and Soviet research that microwave radiation interferes with metabolism, the process by which food is converted to energy and new cells. Besides cataracts, other symptoms of overexposure to microwaves include: headaches, eye strain, fatigue, dizziness, moodiness, irritability, impaired judgement, nervous tension, muscular weakness, a pulling sensation in the scalp, and hair loss, muscle and heart pain, breathing difficulties, and increased perspiration of arms and legs. Some studies indicate that microwave radiation may be cumulative and build up with repeated exposure. No one knows if there is a safe threshold.[39] Paul Brodeur wrote a book on this topic called ***The Zapping of America: Microwaves, their Deadly Risk, the Coverup.***

I used to believe that only a few microwave ovens leaked until I read Dr. Robert Becker's book, ***The Body Electric***. He stated (page 309) that all microwave ovens emit an average of 120 microwatts near the door and many emit higher amounts. He also said that a worn seal can increase user exposure to well over the 5,000 microwatt level, according to the Consumers Union tests. My motto is, if I can't heat it up on the stove, I eat it cold!

Even if microwave ovens did not leak, I am not convinced that eating microwaved-cooked food is nutritionally wise. Look for studies in this area. Until I know with certainty, I have removed my children from the guinea pig list and do not have a microwave oven.

Pole mounted transformer and high current primary lines, a house where you don't want to live!

High voltage line going to substation

CHAPTER 9: VIDEO DISPLAY TERMINALS - A LITTLE BIT OF EVERYTHING FROM THE ELECTROMAGNETIC SPECTRUM

VDT's and TV's are low voltage x-ray machines containing cathode ray tubes. I mentioned earlier that VDT's emit a whole gamut of radiation from high frequency, short wavelength soft x-rays, ultraviolet radiation through the visible spectrum (light) down to decreasing frequencies of longer wavelength which includes infrared, microwaves, radiowaves, and VLF and ELF electromagnetic radiation. Knowing the health effects of VDT radiation and protecting yourself from them is becoming increasingly important as computers become more popular. From 1980 to 1988 the number of VDT's increased from five million to forty million, or roughly three million per year. How many VDT users are women and children? How many female VDT users are pregnant?

Introductory notes

The following information is far from complete and is full of controversy and unanswered questions. Even the expert researchers are divided in opinion. Like many modern appliances, VDT's emit electromagnetic radiation, so why am I emphasizing VDT's? Because users sit twelve to eighteen inches in front of them for hours a day. There are many radiation protection screens on the market today but do not think that the problem has been solved. We have not yet really defined the problem in terms of which type of radiation causes the most damage or the combined effects of several types of radiation.

To understand the problem, you have to know that the electromagnetic emissions generated from VDT's consist of two components, electric and magnetic. Electric currents generate magnetic fields. And, experts disagree as to which field causes the most harm. There has been a lot of research on the adverse health effects of electromagnetic fields and some researchers believe that both magnetic and electric fields are health hazards. Others believe that magnetic fields are more harmful than electric fields and some believe that electric fields are harmless.

Today, no one has shown that electric fields are harmless. The fact that currently available VDT screens (which shield the electric fields generated by computer monitors) considerably reduce computer user complaints is evidence of the adverse health effects of chronic exposure to electric fields.

In a very interesting article printed in *Macworld* (July 1990), Paul Brodeur briefly traced the history of discoveries that computer monitors emit significant radiation in the form of pulsed ELF (extremely low frequency) and VLF (very low frequency) electric and magnetic fields as well as several other types of electromagnetic emissions discussed in detail below. Brodeur says that most of the VDT emissions are pulsed VLF electric and magnetic fields of between 15 and 20 kHz and pulsed ELF electric and magnetic fields of 60 Hz. He also says that the dominant emission given off by VDT's is the ELF magnetic field.

Today, in response to the health problems observed from electromagnetic radiations emitted from computer terminals, many antiglare-antiradiation screens have been developed and are currently available commercially. But, here's the rub: there is no screen available today that blocks the ELF and VLF magnetic fields; only the electric and static fields are blocked. I do not believe that this is enough. Louis Slesin, editor of *VDT News*, says that *"it's a low-frequency-magnetic field problem and screens don't take care of the magnetic field."*

Make no mistake, at common user distances of 12 to 18 inches, the magnetic radiation will permeate the user. For example, Brodeur mentions in the *MacWorld* article that 60 Hz magnetic fields greater than 2 milligauss intensity have been measured at distances of 12 inches from VDT's and fields of about one milligauss at 20 inches from VDT screens. So it's my guess that the user would have to be at least 30 inches away from the screen to avoid getting zapped by magnetic radiation. Other researchers suggest that the user stay arm's length (with extended fingers) distance from their computer, an easier task for the far-sighted than the near-sighted user!

With the above in mind, I will give you more information on the emissions from VDT's which have been gathered from animal studies and from human guinea pigs.

Hard (ionizing) x-rays

In addition to non-ionizing radiation, some TV's leak hard x-rays, especially the 1970 vintage TV's. Based upon a survey of

11,000 homes, about one-half of the twenty five million 1970 TV's sold are still being used. A recent federal survey found that one in twelve VDT's leaked enough x-rays to expose its user to the equivalent of one chest x-ray every other day.[40a] How do you know if you are one of the twelve? Until recently, this question could not be answered. Now, there are several new companies that will measure radiation emissions in your home or workplace. I will briefly discuss this in Chapter 23.

In a 1978 FDA study on VDT radiation, "Evaluation of Radiation," FDA #81-8153, now listed as out of print, the FDA tested 125 monitors from 15 VDT manufacturers. Thirty-three manufacturers chose not to be evaluated. The agency found that ten out of 125 monitors emitted high levels of x-radiation (0.44 - 2 millirems per hour) when the line voltage (from the plug in the wall, usually 120 volts) was raised to 130 volts. Increasing the line voltage is common during a stress test to measure how machines perform as they age. The current FDA standard for the "permissible" amount of radiation emitted from a VDT screen is 0.5 millirads of x-ray at a distance of 6 cm from the screen. This translates to 0.2 millirads per hour from the usual operator distance from the screen of 50 cm (about 18 inches). Since radiation is cumulative, it would take only sixty hours of exposure to accumulate 30 millirads, the amount of one average chest x-ray. [40b]

Let's assume that you are getting the Federal limit of radiation from your VDT of 0.5 millirems/hr and that you work at your VDT six hours per day times five days per week times 50 weeks per year to give a total of 1500 exposure hours. Assuming that one x-ray delivers 30 millirems, this would give you 1500 exposure hours times 0.5 millirems/hour = 750 millirems/year/ 30 millirems per x-ray gives you the equivalent of 25 chest x-rays per year. Do you think that this is safe? *The Toronto Star* reported on high radiation readings found in 11 out of 339 VDT's tested even though they were relatively new machines. [41]

In a 1982 study by C.R. Hirning and J.H. Aitken of the Canadian Government's Radiation Protection Service, Ministry of Labour, the authors, while dismissing any possibility of fetal damage in pregnant VDT operators, admitted that "significant annual doses...may be received by two radiosensitive organs: the female breast (85 millirads) and the thyroid gland (190 millirads)." [42]

Very Low and Extremely Low Frequency Radiation, VLF, ELF

As mentioned in the introductory notes, most VDT radiation is pulsed VLF electromagnetic fields between 15-20 kHz and pulsed ELF radiation of 60 Hz. In particular, the VLF electromagnetic radiation is pulsed rectangular waves from the flyback transformer and horizontal deflection coil in the cathode ray tube (CRT). The ELF electromagnetic radiation is mainly produced by the CRT's vertical deflection coil. Pulsed radiation is biologically active (affects living organisms) as opposed to continuous radiation which is not as active. Also, different shaped pulsed waves have different biological effects. They cannot be predicted. The particular pulsed VLF and ELF waves emitted from VDT's cause multiple biological effects including altered cell growth and development, altered metabolism of protein, fat, carbohydrates and nucleic acids, altered endocrine function leading to decreased neurotransmitters and resulting increased depression and increased suicide, decreased immune system response, morphological changes (teratogenic and developmental) and increased incidence of cancer and leukemia.

Paul Brodeur described a Finnish study from the *Scandinavian Journal of Work, Environment and Health* in 1986 which reported magnetic field intensities from a 50-60 Hz ELF field of eight milligauss at an operator level of twelve inches from the VDT screen and four milligauss at fifteen inches from the screen.[37] In the above mentioned July 1990 *Macworld* article, Brodeur reported similar studies from Canada. You will recall Wertheimer's observation of power line magnetic fields described earlier in which magnetic field strengths of only two-three milligauss found in homes near high current electric wires resulted in increased leukemia rates in the children living there.

Microwaves

We are familiar with microwave radiation from microwave ovens, microwave radios and radar stations. Despite denials from the computer industry, VDT's also emit microwave radiation, which like other sources of microwave radiation, cause cataracts in VDT users. Research on microwave-induced cataracts has been available since the 1960's, largely the effort of Dr. Milton Zaret, an ophthalmologist in Scarsdale, New York. Dr. Zaret has been studying the effects of microwave radiation on the human eye for twenty-five years and discovered that cataracts formed in VDT operators were similar to those he had found in radar operators in the 1960's, where they were apparently the result of microwave exposure. As discussed elsewhere, radiation-induced cataracts occur on the surface of the lens and differ from cataracts associated with aging or heat which occur in the substance of the lens itself.[40,43,44,45]

Mothers beware! VDT's and colored televisions used as computer monitors can also lead to nearsightedness in children, according to many eye doctors.[41] Don't let your children watch TV with their eyeballs glued to the set. And if you or your children use computers, do yourself or your child a healthy favor - buy a shield for your VDT. They are inexpensive compared to dim vision.

Static Electric Fields

In addition to the electromagnetic radiation, VDT screens are surrounded by static electricity, which is a positively charged field. This field neutralizes itself by drawing negative ions from the air around the computer screen, resulting in an increased concentration of positive ions in the air right next to the computer user's face. A high concentration of positive ions in the air or atmosphere is associated with many bad health effects: fatigue, irritability, headaches, lung problems and other metabolic problems, facial rashes, etc. Included in these positively charged particles are dust, cigarette smoke, and chemical pollutants. These positive particles seek out negative charges, usually on the VDT operator's face and this is what creates skin problems including rashes, itchy eyes and dry skin.[40,47,48]

Studies by Dr. John Ott

Dr. John Ott discovered that being in close proximity to a VDT will cause rouleaux clumping of the red blood cells. They line up like a lot of little magnets in a straight line. When he placed a microscope slide of red blood cells in front of a VDT for only five minutes, this clumping action occurred. Ott suggested that this type of clumping is a result of the iron in the hemoglobin becoming magnetized by the magnetic fields from the VDT. If you go to bed after your time on the VDT without any exposure to ultraviolet light (e.g., sunshine), your RBC's will still be "clumped" when you wake up the next morning. However, exposing your body to UV light (sunshine) or to full spectrum radiation-shielded Vita Lite (the Ott light fixture) causes the red blood cells to unclump. How long does this take? Ott reported that a five minute exposure in front of the ultraviolet light source (full spectrum or sunlight) caused the red blood cells on his slide to unclump. My advice is to take frequent breaks in the sunshine and/or put full spectrum lights in your office. Your body needs ultraviolet light for calcium metabolism as well as the "unclumping" of your RBC's [49,50]

Ott did another very interesting experiment. He put some bean plants in front of some old and new TV's and some VDT's to see how effective the new improved models were under ordinary but realistic operating conditions. His results are reported in the *International Journal of Biosocial Research*[51] and *Health Consciousness Magazine.*[50] Plants near the VDT began to show a mottling effect of many small brown spots about 4-5 days after each new leaf opened. The spots increased in size as the leaf developed. Within one week, these leaves turned yellow and dropped off the plant. If the plants were moved away from the VDT, new healthy leaves would appear within one week. If the plants were returned to the proximity of the VDT, the same type of mottling would appear 3-4 days later. There is a picture of one of these leaves in the article. I am sure you would be motivated to get yourself a shield if you saw it.

Dolores Mason wrote a letter to Dr. Ott which documents medical problems of VDT users at Indiana Bell. A copy of this letter is published in *Health Consciousness Magazine.*[50] She describes health problems in 69 out of 110 VDT users at Indiana Bell, including herself. Ms. Mason lists 25 medical conditions including musculo-skeletal problems, vision problems, hypertension, female problems (leading to hysterectomies), throat and thyroid problems, tumors (breast, arms, brain, vagina), nausea, severe headaches, rashes, head and scalp burns, respiratory problems , dizziness and loss of balance. One mother bore a child with burns across the back of his neck. [52,37]

VDT Screens for Protection

There are companies manufacturing many kinds of VDT screens to protect you from some but not all of the radiation I have described. I advise getting a screen which has not only nickel or copper to block the electromagnetic radiation but also lead to block the soft x-rays, plus a grounding wire to neutralize the positive ions. Some companies only use lead. Others only use nickel or copper. Ecosystem (A) manufactures the SomaScreen which is grounded and contains both lead and nickel. The No Rad Screen contains nickel and copper but no lead. Remember that none of these screens block low level (ELF and VLF) magnetic radiation. They only block the VLF and ELF electric field radiation.

In the July 1990 *Macworld* article there is a page on "Rating Radiation Screens." You can keep up with the latest VDT information by reading VDT News edited by Louis Slesin.[53] You can also contact organizations such as the Labor Occupational Health Program (LOHP) in Berkeley, California (A).

The Future: Low Radiation Screens

Liquid crystal monitors, which do not emit radiation, are now available (A). In the near future, video display terminals will be designed with balanced wiring configurations which will cancel out most of the radiation emitted from the current VDT's. The technology for this has been long known and in fact is used routinely in the wiring configurations of ships, because the results of having unbalanced electromagnetic fields would lead to the metal of the ship dissolving via electrolysis and to interference with the magnetic compass which guides the ship. Thus the ship would either sink or get lost!

It's only a matter of time before magnetically balanced VDT's will be available. IBM has already announced their intent to do this.[54]

A Note of Caution for Pregnant Women

Because currently available VDT screens do not block VLF and ELF magnetic radiation, which has been shown to double the rate of fetal deaths in animal studies.[55] I do not recommend that pregnant women use computers (other than the liquid crystal monitor described above). Who wants to take a chance on loosing their child or having an abnormal child ? Until we are sure that we are protected, it is better to err on the side of safety than to use shields which only partially block radiation.

CHAPTER 10:
ULTRAVIOLET RADIATION, MAL-ILLUMINATION AND FULL-SPECTRUM LIGHT

Ultraviolet Radiation

Light is a nutritional substance required for health. Sunshine, containing ultraviolet light is necessary for synthesis of vitamin D in vivo (inside the body) which is required for the metabolism of calcium and phosphorus. It has been estimated that exposing the body in a bathing suit to the midday sun for five minutes will provide the equivalent of 200-300 i.u. of vitamin D. However, excessive tanning, including tanning booths, can cause increased risk to skin cancer and immune system suppression (decreased resistance to disease and infection). The incidence of skin cancer has increased five times in the last 20 years and about 90% of this is believed to be caused by exposure to ultraviolet radiation from the sun.

The increased incidence of skin cancer is due to the increased popularity of sunbathing and a gradual decrease in the ozone layer, our most protective sunscreen. Recently, unexplained loss of ozone above Antarctica has prompted a reevaluation of the chlorofluorocarbon (CFC) hazard by the EPA. CFC's are widely used as refrigerator coolants, in auto air conditioners, as foaming agents and in aerosol sprays.[4]

The National Academy of Sciences estimates that a 1% decrease in the ozone layer will cause a 4-10% increase in skin cancer. A 10% depletion will cause at least a 40% increase in skin cancer. When this occurs, sunscreen lotions will be ineffective.

CFC's are not the only man-made cause of decreased ozone. For example, cars consume oxygen and produce carbon monoxide to the tune of 18.5 pounds of oxygen per gallon of gasoline. Scientists have discovered a carbon monoxide cloud in the upper stratosphere weighing 240 million metric tons. Each pollutant, whether it be carbon monoxide, nitrogen oxides, sulfur oxides, methane gas or CFC's depletes oxygen from the ozone layer, using up oxygen at the rate of 13.5 trillion metric tons per year. This, in turn, depletes the ozone shield. Today our ozone shield is less than three miles thick. In 1954, it was 9.4 miles thick.[56]

Because two-thirds of our protective ozone shield has been depleted in the last 33 years, intense ultraviolet radiation is damaging oxygen-producing plankton and microorganisms living in the ocean. Even if we stopped all our pollution today, it would take seven years for the pollution present today to reach the ozone shield where it would continue to destroy the ozone for another 120 years.

Add to this loss the disappearance of our rain forests where we have cut 80% of the trees in the last 200 years. The last great forest, the Amazon is now being burned at the rate of one football field per second, or 13 million acres in 1989, which is equivalent to 13 million times 100 tons of oxygen lost.[56]

Add to this the acid rain created from the mixture of air pollutants and moisture which kills plants and trees. Again, more oxygen depletion.

Full Spectrum Lighting

Man was meant to be exposed to the full spectrum of light. You can read all about this super nutrient in John Ott's books, **Light, Radiation and You; Health and Light**, and others, his newsletter [57], and numerous articles in the journals.[49,50,51,58]

Like food, full spectrum light is a nutritional substance. Most commercial lights such as incandescent bulbs and all fluorescent lights except those designated as full-spectrum (Ott Light, Vita Lite, etc.) do not produce full-spectrum light and therefore induce nutritional deficiencies. Full spectrum light must enter through the eyes to influence the pituitary and pineal glands which affect the entire endocrine system. Thus, wearing sunglasses or tinted contact lenses can rob you of energy and health by preventing the blocked light wavelengths from entering your endocrine system via the pineal and pituitary glands.

The pineal gland secretes melatonin, a hormone that controls many bodily functions including those related to sleep, ovulation and the secretion of other hormones. The pituitary has a profound effect in controlling the secretions of the other endocrine organs, including the ovaries, adrenals and thyroid glands.

The effects of mal-illumination resulting from artificial lights have been fully documented: hyperactivity, irritability, attention deficit and nervous fatigue in school children; reduced productivity and fatigue in office workers; and increased cavities, increased risk of osteoporosis (because of decreased absorption of calcium), and depression in the general public.[58,59,60] Even worse, Valerie Beral et al. found a doubling of malignant melanoma in workers exposed to artificial (not full-spectrum) fluorescent lighting in the workplace compared to non-exposed workers.[61,62]

Russian researchers have documented that the body's tolerance to environmental pollutants is increased by full-spectrum light. West Germany restricts the use of cool white limited-spectrum fluorescent lights in public buildings because of their adverse health effects.

You can buy full-spectrum ultraviolet transmitting eyeglasses from Younger Optics in Los Angeles and full-spectrum contact lenses from the Wesley-Jessen Company in Chicago, Illinois (A).

You can also buy incandescent full spectrum lights. These lights are manufactured in the Netherlands and available from Baubiologie Hardware (A). These new lights do not have the disadvantages of the full spectrum fluorescent tubes (radiation leakage and flickering), and can be screwed into any light socket.

Reprinted by permission of NEA, Inc.

"We take the view that if you can't stand the pollution, you should stay out of the environment."

SECTION II:
HOW TO PROTECT YOURSELF FROM THE EFFECTS OF RADIATION

In this section are things to do, things to avoid, foods to eat and supplements which offer some measure of protection against radiation and other sources of free radical pathology. Remember, the processes of aging, cancer, heart disease, inflammatory diseases (arthritis, bronchitis, cystitis, rhinitis, sinusitis, etc.), allergies and other chronic degenerative diseases have the common thread of free radical pathology running through them. It will help to know about foods and nutritional substances which will quench the free radical, the nasty little critter with the unpaired electron that initiates the free radical chain reaction. There are many supplements, such as antioxidants which not only stimulate the immune system, but also stop free radical reactions.

Author unknown

"Gentlemen, this is a test. Chemicals
you certified as safe are being
circulated in your area..."

CHAPTER 11: BALANCE YOUR BODY CHEMISTRY BY RESTORING A HEALTHY RELATIONSHIP WITH NATURE AND ALL CREATURES OF THE EARTH.

What makes our bodies get out of balance? How did we mess up our physiology, develop enzyme, vitamin, mineral and other nutrient deficiencies and set the stage for chronic degenerative diseases? In our technological evolution, we have lost our contact with nature and her rhythms. Sometimes we even believe that we are better than nature, that we can manufacture better foods, better organs, better everything.

Before the commercialization of agriculture, we maintained a harmonious cyclical relationship with the earth and its creatures. Not a single thing was wasted. The animals ate what man did not eat. Manure from the animals was mixed with green cuttings, weeds, egg shells, dirt and water, and magically transformed into a beautiful compost. This natural compost nourishes the soil so that year after year, the soil improves in quality. You can taste this quality in organically grown fruits and vegetables and in the animal products from animals who are treated with love and are fed organic foods. People who eat this food don't get fat. It's hard to overeat foods filled with so much nutrition. Animals who are raised humanely, fed organic foods, and who are allowed to walk uncrowded on the earth, do not develop the diseases observed in commercially raised animals.

Compare this to our linear, high-tech modern life style. It's opposite to the natural cycle of life. It's a one-way flow from the factory or store into the garbage or waste container. Little reverence or respect. Lots of consumption, pollution and depletion of the earth.

Compare the soil of organic farming to the soil of agribusiness farmers. It's chemically fertilized, mechanically plowed and mono-cropped by agribusinessmen in air-conditioned farm equipment. Commercial food tastes dead compared to organic food because it is dead, having been grown in dead soil. And look what's happening to the dead soil. It's disappearing - drying up, blowing away, and becoming a growing wasteland.

Since we cannot separate ourselves or our animals from the earth, whatever we have done to the earth and its inhabitants is now being done to us. We have polluted the earth with toxic chemicals, pesticides, disease organisms from so-called disposable diapers and other toxic waste materials to antibiotic-resistant pathogens, and radiation - the ultimate toxic waste. Chernobyl will be radioactive for as long as the earth lives.

The American Indians understood the importance of maintaining a harmless, harmonious relationship with the earth and its creatures large and small. We came to America, took the land, destroyed its beautiful vegetation, depleted the soil and killed many of the animals and the Indians. The commercialization of the beef and poultry industry separated these animals from earth and thus from their contact with nature. Instead of treating them with love and kindness and allowing them to have a harmonious contact with nature, we have drugged them and filled them with pesticides, antibiotics, hormones and tranquilizers. Because of this, they have become weakened and diseased. We eat these diseased, drugged animals and in turn have become diseased and drugged. What we have done to the earth and its creatures is now being done to us because we cannot separate ourselves from the earth and its creatures. This was eloquently described in the 1800's by Chief Seattle, a great American Indian.

"I will make one condition
The white man must treat the beasts of this land
as his brothers.
For whatever happens to the beasts
soon happens to man.
All things are connected.
One thing we know:
Our God is the same.
This earth is precious to Him...
This we know:
The earth does not belong to man:
Man belongs to the earth.
This we know:
All things are connected
Like the blood which unites one family.
All things are connected.
Whatever befalls the earth
Befalls the sons of the earth.
Man did not weave the web of life.
He is merely a strand in it.
Whatever he does to the web,
He does to himself."

I took the above quotation from page 381 of John Robbin's book, ***Diet for a New America,*** but I have seen it in numerous other publications. Read Robbin's book for a deeply moving narration of the plight of the earth and its animals.

So now you know the answer to my question. How do we balance the body chemistry? It is more than just taking vitamins, minerals and enzymes. We must establish reverence and respect for the earth and its creatures, because whatever we do to help the earth and its creatures will help us. Whatever we lack, we have created by being out of balance with the earth; by eating dead, devitalized, commercially processed foods devoid of life and filled with disease organisms, drugs, hormones, pesticides and radiation products. You can take vitamin A if you lack it but do you really think that this is the answer to your health? Do you think that chemical labs have the answer to health? Instead, seek a more organic way of life. Support the organic and biodynamic food industry. Eat more whole foods and less junk foods. I will describe this in detail below.

Aside from our genetic predispositions, organ weaknesses arise from eating dead, depleted, diseased foods. A healthy, well-nourished organ is less susceptible to free radical damage than a weak, unbalanced organ. Chronic degenerative conditions can be asymptomatic until the medical crisis care stage. Either we choose drugs which control the body chemistry or we provide the nutrition which allows the body to control itself. Which do you prefer?

I am not saying that drugs don't work. They do, but their effect is uni-directional. They do not modulate your blood chemistry, i.e. raise or lower your glucose, cholesterol, etc. to maintain homeostasis. For example, taking insulin lowers the blood sugar but the dosage must be carefully monitored because the blood sugar does not stop dropping at homeostasis (the optimum range required to maintain the life of the organism). Also, synthetic drugs can have terrible side effects, sometimes worse than the condition for which they are being used.

Foods work differently than drugs. If whole foods are eaten and properly digested, absorbed and assimilated, homeostasis will be achieved and maintained. In other words, nutritional substances have **modulating** effects. For example, through complex metabolic processes, the blood sugar is raised or lowered as needed to maintain optimum blood glucose levels. It amazes me that we think we can compete with these complex processes by eating synthetic substances.

Balance your body chemistry. Find out what enzymes and whole foods you need through the proper testing: blood, serum, urine chemistry, hair analysis, muscle testing, etc. Use whatever method you find best for yourself. If you can balance your body chemistry so that each cell has the optimum amount of nutrients, you will stay well. If you are sick, a self-healing process will occur. Healing is a self-initiated process.

"Let me get this straight. We plant the chemically treated seeds in the artificially fertilized soil... Then we spray 'em from the air... And **then** we irradiate them?!?"...

CHAPTER 12: LIMIT, MINIMIZE, AVOID RADIATION

Find out what sources of radiation are common in your life and minimize your proximity to them.

1. **Keep a healthy distance from your TV (at least six feet).** If you need to glue your eyeballs on your screen, get some glasses. You need them.

2. **Get a VDT screen.** See appendix A for addresses.

3. **Be careful of microwave ovens.** All of them leak. If you have to use one, don't stand hypnotized in front of it watching the food bubble. Remember, you will not feel its heat in your body. The only thing you will notice is the cataract and by then it will be too late.

4. **Don't live or work near strong electromagnetic fields.** It's a matter of choice - yours .

5. **Use full-spectrum lights**, for example, Vita Lites, patented by Dr. John Ott, who studied the effects of mal-illumination. These effects, which include decreased calcium absorption resulting in increased cavities, hyperactivity in children, learning disabilities, and malignant melanoma, are discussed in Chapter 10. [40,50,51,57-62]

6. **Check your bedroom.** It is one of the potentially worst places for electromagnetic radiation if it contains TV's, dial face clocks, electric blankets, water beds, clock radios, etc. Eliminate or buy magnetically balanced water bed heaters. Move all other appliances away from your bed, at least three feet.

7. **Check your house for radiation.** There are several companies which do this. See Safe Environments (A).

Reduce your exposure to carcinogens in food, air and water.

1. **Wash vegetables and fruits** to minimize insecticides, pesticides and other chemicals. You can use a vinegar wash or a detergent wash, followed by a rinse.

Hydrogen peroxide bath: Some people prefer to use 3% hydrogen peroxide (1/4 cup in a sink full of cold water). Thin skinned vegetables are soaked 20 minutes, thick skinned ones 30 minutes.

2. Avoid rancid fats and moldy foods. Don't heat unsaturated oils - this promotes peroxidation. Instead, cook with olive oil (which has only one double bond and is more stable than polyunsaturated oils), canola oil or butter. Secondly, some molds are unhealthy. For example, peanuts and other foods develop a carcinogenic mold (aflatoxin) which has been shown to cause cancer in animals. Since each species requires a different amount, no one can say how much causes cancer in humans.[63] So, use fresh peanuts, or limit your use, and try other nut butters such as almond, cashew or sesame seed butter (tahini). All are excellent sources of essential fatty acids and sesame seeds have definite, positive effects on the immune system (see section on foods).

3. Keep your colon healthy. Otherwise, the body will harbor many toxins, including ones that cause cancer. How to avoid constipation? Eat a high complex carbohydrate, high fiber, low sugar, low fat diet. Don't take laxatives. Instead, eat a natural foods diet and take the food enzymes discussed in Chapter 14.

4. Get some sunlight everyday. I don't mean to cook yourself. Fifteen minutes daily is fine. Sunlight is essential for life and for the production of vitamin D which is required for calcium absorption. However, avoid overexposure (e.g. dark tans), especially during hours of most intense sunlight (between 10 am and 2 pm) and in tanning booths. Remember, our atmosphere is less shielded from harmful radiation now due to chemically induced destruction of part of the ozone layer.

5. Avoid smoking, either actively or passively. The increase in no-smoking areas in work places, airplanes and recreational areas is making this increasingly easier to achieve. The Environmental Protection Agency will declare second-hand cigarette smoke a known carcinogen and give the first official government estimate that passive smoking causes 3,000 cases of lung cancer among non-smokers each year.[64]

There is mounting evidence of double trouble resulting from weather-tight homes which allow radon, a radioactive gas present naturally in indoor air to accumulate in high concentrations. In smoke-filled rooms that are air-tight, the radioactive decay products of radon become attached to airborne tar particles in cigarette smoke. These particles are inhaled by smokers and nonsmokers and are deposited in high concentrations in extremely small areas of the lung which then undergoes cellular changes from the radiation. The EPA estimates that up to 20,000 Americans may be dying each year from radon-induced lung cancer.

Although radon presents a real cancer risk, there are researchers such as Gofman and Sternglass who believe that the federal government is using concern about radon, a natural form of radiation, as a method to deflect concern about man-made sources of radiation.[65] For example, Michael Lafavore used a lot of research from the EPA and DOE to write his book, ***Radon, the Invisible Threat***. But many years of government research on man-made radiation is still not available or is difficult to obtain.

6. Drink distilled water. Which is best: spring water or distilled water? I recommend distilled water that has been filtered through carbon and ozonated. It took me ten years to come to the conclusion that distilled water is best. Why? For several reasons. My daughter, now ten, has dental fluorosis (mottled teeth) from drinking pure spring water - she has never had any other source of fluoride, including dental work or tooth paste. Pure spring water is increasingly hard to obtain due to ground-water contamination by chemicals and toxic wastes from the chemical industry.

If you buy spring water, check the quality of the water you buy. Ask for a current assay (a list of all substances in the water). If you can't get one, beware of the water company. You may be buying your own tap water (or worse)!

If you use a water filter, no one can promise how long it will last because they cannot know how polluted your tap water is unless they analyze it. The more polluted the water, the shorter the life of the filter.

Secondly, no carbon filter can remove fluoride, which I consider to be one of the worst toxins. Please see more information below.

Thirdly, I believe that pure distilled water is the best for health and healing. I have observed that some people who drink hard water have traces of protein in their urine (not good). I believe that they are also at increased risk of cardiovascular disease. I have no reference for this except the research of Dr. Howard Loomis of 21st Century Nutrition who has spent the last ten years correlating diet and nutrition with a special 24-hour urinalysis which he has developed. This is described in Chapter 14 on enzyme nutrition.

Here is a list of unwanted things in your water supply:

Pathogenic bacteria that cause diseases like dysentery, hepatitis, typhoid, cholera and "stomach flu."

Synthetic organic chemicals such as insecticides and their residues, and products of the petrochemical industry. Familiar names such as carbon tetrachloride and other chlorinated hydrocarbons, benzene and its derivatives and any one or more of over 12,000 industrial chemicals.

Added chemicals such as chlorine and fluorine. Chlorine and by-products of chlorine combined with organic matter in the water, such as chloroform, are associated with arteriosclerosis, cancer, premature senility and decreased life span. More on this may be found in books on water by Drs. K.W. Donsbach and Paul Bragg (B).

Fluoride, a cumulative enzyme poison has so many pathological effects that it has been called, ***The Aging Factor,*** in John Yiamouyiannis' book entitled the same. Dr. Yiamouyiannis has spent many years gathering irrefutable evidence that proves fluoridation does not prevent cavities. To do this he analyzed the largest data base ever gathered (by the National Institute of Dental Research of the U.S. Public Health Service) to determine whether there is a relationship between fluoridation and tooth decay. This nationwide survey examined 39,207 U.S. school children aged 5-17 from 84 geographical areas. The decay-free children in the fluoridated, non-fluoridated and partially fluoridated areas was 34%, 35% and 31%, respectively. In other words, 70% of our children have decayed teeth regardless of fluoridation! Yet we read claims that our children have less tooth decay now than before fluoridation! Yiamouyiannis' work is available from his organization, the Safe Water Foundation (A). [66-69]

Here's another good one! I read this title in the *San Jose Mercury* newspaper (January 29, 1990): "Fluoride found to cause (bone) cancer in rats; human risk discounted."[70] Don't humans have bones too? How can anyone believe that humans can ingest a carcinogenic substance daily without risk of getting cancer?

Dr. George L. Waldbott, who wrote ***Fluoridation, the Great Dilemma***, and other authors including Donsbach and Russell-Manning (B) have reported many other effects from drinking fluoridated water including aching bones, fatigue, brown mottled teeth, stiff joints, loss of appetite, nausea, vomiting, etc. It is ironic that fluoride, which is supposed to make your bones and teeth healthy can cause osteosclerosis (too much calcification) and osteomalacia (adult rickets).

Heavy metals (lead, cadmium, arsenic, mercury, nickel, etc.) compete with nutrient minerals in the body and cause a wide range of health problems (cancer, nervous system disorders, etc.) as well as violent behavior.[44]

7. Avoid silver amalgam fillings. Why? Hal Huggins,[45] Betsy Russell-Manning (B), Roy Kupsinel and others will tell you in well-documented information from many clinical studies. The trouble here is caused by the mercury (and other metals) in the amalgam (up to 50%) which leak into the body and wreak havoc.

A recent research study will lay permanently to rest the erroneous position of members of the American Dental Association that the mercury does not leak out of the filling into the body. In this study, radioactive mercury (Hg 203) was placed in adult sheep teeth and was found to appear in various organs and tissues within 29 days. Evidence of mercury uptake determined by whole-body scanning revealed three uptake sites: lungs, gastrointestinal and jaw tissue. Once absorbed at these sites, high concentrations of mercury rapidly localized in the kidneys and the liver.[71] This study explains my observation of kidney and bladder problems, including kidney infections and cystitis, in women who use contraceptive vaginal foams and gels , all of which contain mercury.

The pathology of mercury poisoning is complicated because mercury binds to many different sites in the body. For example, mercury is especially attracted to sulfur and there are many "sulfur sites" in the body. Sulfur is found in red blood cells, colinergic receptors in cardiac muscle, and in the mitotic spindle of the cell, the seat of genetic reproduction. Mercury will poison any enzymes containing sulfur such as those involved in the formation of insulin, the coagulation of blood, and the biosynthesis of DNA precursors.

The list of signs and symptoms of mercury toxicity is a long one. Dr. Roy Kupsinel, M.D., has a magazine called *Health Consciousness*, which devotes a large section in each issue to research on mercury-amalgam sensitivity.[72]

Mercury has the curious talent of being able to mimic other diseases which have not previously been associated with this metal, such as multiple sclerosis, rheumatoid arthritis, arthritis, scleroderma and lupus erythematosis. If mercury is the causal factor and the damage is not irreversible, removal of the mercury will initiate healing. There are many clinical reports of people jumping out of their wheel chairs and leaping out of their beds following removal of their amalgams and replacement with gold or composites.

You should know some of the signs and symptoms associated with mercury sensitivity (be it from amalgams, fish, or contraceptive creams and gels). Just to mention a few categories: nervous system (headaches, migraines, convulsions, ringing in ears, memory loss, irritability, tremors); cardiovascular system (irregular, fast or slow heart beat, high or low blood pressure, chest pain); skin (allergies, eczema); connective tissue (arthritis, and the diseases listed above such as scleroderma); gastrointestinal (abdominal cramps, bowel problems such as colitis); respiratory (persistent cough, emphysema) and many others (B).[72]

Organic mercury, called methyl mercury is the most toxic form, 100 times more toxic than elemental mercury. You can make your own methyl mercury in vivo just by having dental amalgams plus methyl donating organisms. Mercury is methylated by not only Candida albicans but by many other organisms - 40 strains of Streptococcus, 30 strains of Staphylococcus, E. coli, and others, according to research by Dr. Hal Huggins, spokesperson for The Toxic Element Research Foundation, TERF (A). Copies of Dr. Huggins' research papers on methyl mercury are available from TERF for $25.

Huggins also reports that mercury can transform a perfectly normal bacteria into one that is resistant to antibiotics. Also, there is evidence of a direct effect of silver amalgams on the white blood cells. In 1984, Dr. D. Eggleston, D.D.S. reported a decrease in the helper/suppressor (T4/T8) ratio in patients receiving silver amalgams which reverted to normal upon their removal.[73] A similar but usually irreversible pattern develops in AIDS patients.

I am frequently asked why not everyone has symptoms of mercury sensitivity. There is no clear cut answer to this question but I believe it is related to each person's ability to bind mercury to substances which are then excreted. For example, certain amino acids (methionine and cysteine) contain sulfur as sulfhydryl groups. Substances that bind mercury and other metals are called chelating agents and this principle of chelation is used in intravenous treatments to remove heavy metal toxins (chelation therapy). In my experience, the effects of mercury toxicity do not necessarily parallel the amount found in a hair analysis. I have seen very sick people with small amounts of mercury and asymptomatic people with much larger amounts. The former group are extremely hypersensitive to mercury, and may have limited methods of excreting it.

Let me end this section with a cheerful note! There are several methods of mercury detoxification. If you decide to have your amalgams replaced with gold, porcelain or composites, choose a dentist who is trained to do it, otherwise your symptoms can become much worse. Secondly, there are nutritional supplements which aid removal of mercury and other heavy metals: wheat or barley grass, vitamin C (helps solubilize metals, acts as chelating agent), charcoal (helps bind mercury), and selenium-containing substances (glutathione, garlic).

"Seriously, They're the only ones laying any eggs around here."

CHAPTER 13: EAT NATURAL, UNPROCESSED FOODS, STAY CLOSE TO MOTHER NATURE

There is comprehensive study published by Cornell University Press which delineates the relationship between diet and the risk of developing disease. The study fills 920 pages of epidemiology resulting from the study of the dietary habits of 6,500 Chinese people. Let me paraphrase some of the findings in this report:[74]

* Obesity is more related to what than how much people eat. Chinese people eat 20% more calories than Americans but Americans are 25% fatter. The main dietary differences are fat and starch. The Chinese eat only one third as much fat as Americans while eating twice as much starch. The body readily stores fat but burns off a larger proportion of the carbohydrates consumed. Differences attributable to exercise have not yet been reported.

* The Chinese data imply that a maximum of only 20% of calories should be derived from fat, and preferably only 10-15%. So, reducing dietary fat to less than 30% of the calories as is currently recommended for Americans, may not be enough to curb the risk of heart disease and cancer.

* High protein diets, especially animal protein, was also linked to chronic disease. Americans eat one third more protein than the Chinese and 70% of American protein comes from animals, while only seven percent of protein consumed by Chinese does.

* In addition, childhood diets which are high in calories, protein, calcium and fat promote rapid growth early in life, early menses and these may increase a woman's risk of developing cancer of the uterus and breast. Chinese women, who rarely suffer these cancers, start their menses three to six years later than Americans.

* Dietary calcium does not prevent osteoporosis according to the study. Most Chinese consume no dairy products and get all of their calcium from vegetables. Although they consume only one half the calcium that Americans do, osteoporosis is uncommon in China despite an average life expectancy of seventy years. This study was masterminded by Dr. T. Colin Campbell, a nutritional biochemist from Cornell University.

In summary, it appears that a low fat, low protein, high complex carbohydrate diet promotes health and longevity.

Nutrition Rules for Health Maintenance

1. Eat fresh foods first, frozen second, and canned only if there are no alternatives (such as canned-tuna and olives). Support your organic farmer! If you can't find one, use the veggie/fruit wash described on page 53.

2 . Avoid white sugar. The body runs on glucose and it likes to manufacture its own as needed, from complex carbohydrates, fats and protein. Sucrose breaks down into its constituents, glucose and fructose during digestion (providing your body contains some sucrase. Recent evidence described in Dr. Linus Pauling's book, ***How to Live Longer and Feel Better,*** indicates that fructose may be the bad guy in the sucrose molecule because it increases blood fats. White sugar contains no vitamins, especially B-1, B-6 and K. Vitamin B-6 is the anti-cavity factor present in raw sugar. Raw sugar has up to 5% minerals (including magnesium, zinc, copper, manganese and chromium which are required to metabolize the sugar) whereas sucrose has only 0.5% minerals. These are the same minerals that are low in people with hypoglycemia or diabetes and in women with PMS. In fact, sugar does all the metabolic things that have been attributed to eating cholesterol. It decreases HDL cholesterol (the good kind that protects you from heart disease), increases triglycerides, increases uric acid (a risk factor in gout and heart disease), increases platelet adhesiveness (makes your blood cells stick together), not to mention the behavioral effects. All of this research is presented in Pauling's book (B) and in the Wright/Gaby Nutritional Seminars available on cassette tapes (A).

The association of sugar with diabetes is well known. New studies have emerged linking sugar consumption to other diseases. In 1949, Dr. Melvin Page, D.D.S. wrote a book entitled ***Degeneration Regeneration,*** in which he reported a life time of research on the effects of sugar on dental caries and many other conditions including diabetes and cancer.

This research documents evidence linking refined sucrose consumption with cardiovascular disease, cancer in general, breast cancer in particular,[76] and liver damage .[77]

So how do we sweeten our goodies? The best ways are **limited** amounts of raw honey, grade-C maple syrup, unsulphured molasses, and barley malt. Concentrated fruit juices are an excellent way of sweetening baked goods. In defense of fructose, there is some very recent clinical evidence that eating a high fiber food containing fructose will actually lower your blood cholesterol and triglycerides.

There is a new organic form of sugar (sucrose) available called Sucanat which is simply the dried powdered crystals derived from organic sugar cane juice. But remember, just as there is a condition called "lactose" (e.g. milk sugar) intolerance, so also is there a condition called "sucrase" (cane sugar) intolerance. These are enzyme deficiencies. I have seen this cause seizures in at least three of my clients, including a six year old girl. When the sugar is eliminated, so are the seizures, a small price to pay for avoiding mind-paralyzing drugs for life!

3. Avoid synthetic sweeteners. The one I am most concerned about is Aspartame, manufactured by G.D. Searle & Co., and approved for use in foods and beverages in 1983. Aspartame is the major ingredient in NutraSweet and Equal. It is composed of two amino acids, phenylalanine and aspartic acid held together chemically by methanol (wood alcohol). It breaks down into these substances in the gut and can also decompose in 86 degree heat which can occur during storage of foods and beverages containing aspartame.

Methanol toxicity is well known and includes a long list of symptoms such as dizziness, visual impairment, disorientation and loss of equilibrium. How much does it take to produce these symptoms? This is anybody's guess. But when Aspartame-sweetened diet sodas and soft drinks are used to replace fluid loss during exercise in hot climates, the amount of methanol ingested can easily reach 250 mg per day. This is 32 times the EPA's recommended limit of consumption for methanol, a cumulative toxin.

The FDA has received hundreds of complaints since Aspartame's advent into our diet in 1983. They include sleeplessness, dizziness, seizures and hallucinations. Other symptoms listed by the Aspartame Resource Center (ARC, a branch of the Community Nutrition Institute, Washington, D.C.) include: headaches, loss of limb control, blurred vision, loss of depth perception, memory loss, speech difficulty, depression and early menopause.[78-80] In my own practice I have had several clients present with a history of seizures following consumption of beverages containing NutraSweet who were placed on Dilantin, an anti-epileptic drug with severe side effects.

Phenylalanine, one of the two amino acids in Aspartame, is also on the "iffy" list, according to ARC, who says that as many as five million Americans have a genetic intolerance for phenylalanine. This can lead to PKU (phenylketonuria), a disorder that can lead to mental retardation, in offspring born with this defect. Some researchers believe that pregnant women with this genetic intolerance who use Aspartame-sweetened products may produce children with lowered IQ levels. Other areas of possible risks include brain damaging effects similar to those of monosodium glutamate (MSG) and brain-tumor formation.[78-80]

James Turner, lawyer, consumer advocate and author says, *"Aspartame breaks down into 30 to 40 % more products than the FDA has even considered."*[79] So there are many possible chemical causes of all the observed side effects.

My recommendation is: read the label and don't buy it if you see the words Aspartame, NutraSweet or Equal.

What about Saccharin? Have you read the label on packages containing it. Briefly, it says that Saccharin has been found to cause fetal abnormalities and cancer in laboratory animals. What more can I say?

Why does everything have to taste sweet? What's wrong with pure water with a slice of lemon or lime or fruit juice diluted with some crisp, sparkling water? Try it? Give some to your kids, too!

4. Avoid white, enriched and even commercial whole wheat bread. Buy whole wheat bread from small bakeries and sold in your local health food stores. Briefly, white flour (and commercial whole wheat flour) contain flour bleach which forms a substance (alloxan) that causes diabetes in animals.[81] Would you knowingly eat a chemical that destroyed the beta cells of your pancreas - the cells that produce insulin? Do you think there is a connection between the rapid growth of diabetes in America and the use of bleached white flour?

Flour bleach is used not just to bleach flour, but to age it, soften it, and as an insecticide (because bugs won't go near it). Commercial bread also contains other insecticides, such as methyl bromide and other brominated hydrocarbons; rancid oils, due to the water used to separate out the bran and germ, which speeds rancidification of the remaining oil in the flour; wheat softeners, and anti-mold chemicals.

4. Avoid partially-hydrogenated oils and foods containing them (also called margarine and vegetable shortening). This includes many foods (?) such as most commercial breads, crackers, cakes, cookies, artificial creamers, cakes, candy bars, etc. Oils are hydrogenated with nickel catalyst to make them solid. During the process a synthetic or "trans" fat is produced. It is this synthetic fat that causes the damage in your body. The makers of margarine tell you that you should eat margarine because it does not contain cholesterol but I will tell you another story. The synthetic (trans) fat in margarine (up to 40% in some brands) inhibits an enzyme in your liver which converts your cholesterol to bile salts required for digestion. This causes your cholesterol to increase (because it can't be converted to bile), your digestion to decrease, and an increased tendency to gallstones. So beware of partially-hydrogenated oils if you have these tendencies (or even if you don't and don't want to develop them!).

In addition, trans fats interfere with the normal permeability of the cell membrane and with the essential fatty acid (EFA) pathway. So what, you say? Well, from the GLA (gamma linolenic acid, an essential fatty acid) comes PGE-1, an anti-inflammatory prostaglandin, a hormone-like substance essential to the proper functioning of your immune system. A deficiency of PGE-1 can contribute to atopic eczema, PMS (premenstrual syndrome), arthritis, cystic mastitis, prostate problems, tendency to weight gain and elevated blood pressure, to name a few. Another myth is that margarine contains less calories than butter - it doesn't! You can read more about this synthetic food in Dr. Carlton Fredericks' book (B) or listen to Dr. Jeffrey Bland talk about this topic in one of his many nutritional seminars, available on tape (A) and in his books.

Be careful when you read labels (we all read the labels don't we?) because many health conscious companies are printing "non-hydrogenated" on the label to designate that it does not contain partially hydrogenated oils.

5. Avoid homogenized, pasteurized milk. Homogenized milk, according to Dr. Kurt Oster and others, is a cardiotoxin, a direct result of the homogenization process which encapsulates an enzyme in milk fat called xanthine oxidase (XO). In unhomogenized milk, XO is digested in the stomach and small intestine into smaller particles which are used or excreted. In homogenized milk, XO is encapsulated into tiny particles (liposomes) which protect the active XO from stomach acids and digestive enzymes and allow it to pass through the intestinal wall into the lymphatic system and ultimately, the blood stream. There, active XO attacks plasmalogen, a vital structural component of the cell membrane in heart and arterial wall tissue. A harmful free radical, superoxide anion, is produced during this process. Active XO has been isolated from arterial plaques and extensive plasmalogen tissue destruction has been observed in the arteries of heart disease victims. XO converts nucleic acids to uric acid (which you don't need if you have a tendency towards gout!). Folic acid inhibits XO and helps rebuild plasmalogen in addition to decreasing uric acid formation. Also, female sex hormones inhibit XO which is one reason why atherosclerosis is less frequent in women prior to menopause than men (there are other reasons). Nicholas Sampsidis describes Dr. Oster's work in his book on homogenized milk (B).

Now, what's wrong with pasteurized milk? If you really want to know, read Dr. William Campbell Douglass' humorous and well-documented book, ***The Milk of Human Kindness is Not Pasteurized.*** It's guaranteed to make you join the fight for the raw milk industry. (I say 'fight' because raw milk is illegal in 14 states).

What's the difference between raw and pasteurized milk? Raw milk is a living food containing enzymes, protective bacteria (such as lactic acid bacillus), vitamins, minerals, fats, proteins and sugar (lactose), etc. If you leave raw milk at room temperature the lactic acid bacillus will change the lactose into lactic acid. This is called sour milk. It is even better for you than sweet milk. On the other hand, try leaving pasteurized milk out at room temperature. Since pasteurized milk is a dead food, it will rot, just like any dead animal. Pasteurization kills the lactic acid bacillus which turns milk sour. It kills or alters the enzymes, proteins, fats and sugar in milk. Some of these enzymes, such as phosphatase, are essential for calcium absorption and the assimilation of other minerals. Only raw milk and whole grains (containing the wheat germ and bran and no flour bleach) contain phosphatase.

The propaganda that pasteurized milk is safer than raw milk is wrong. In fact, raw milk contains enzymes and antibodies that make it less susceptible to bacterial contamination. However, raw milk and pasteurized milk are **equally** susceptible to man's contamination! The only way to make milk safe is to keep it clean - clean cows, clean dairies. Instead we kill milk by heating it (pasteurization).

Dr. Douglass cites recorded cases to show that when milk gets contaminated, it's usually **after** pasteurization. For example, on page 25 of his book, he states that in 1945 there were 450 reported cases of infections caused by raw milk and 1,490 cases caused by pasteurized milk. There are many other cited examples.

Now let's lay to rest some of the rumors I'm sure you've heard, thanks to Douglass' research. Contrary by what you've been told by the National Dairy Council, the AMA, the Center for Disease Control (CDC), and even perhaps your mother, you can drink raw milk **without** getting tuberculosis even if the cow has TB. Says Dr. Douglass on page 10, "*the blood-brain barrier prevents the tubercule bacteria from passing into the milk. It was tuberculous milkers who infected the milk by coughing into it.*" Neither is brucellosis (undulant fever) contracted from milk but from direct contact with the infected animals. The milker would get undulant fever but seldom his children who drank the milk (page 11).

If you or your children drink raw milk get diarrhea, don't call the CDC. They may tell you to cancel your milk order because the raw milk contains campylobacter, a nasty germ that causes diarrhea. Not so, says Dr. Douglass. He says that campylobacter has never been recovered from raw milk because it contains an enzyme (lactoperoxidase) which inactivates this germ. So where does it come from? Primarily chicken, turkey and tap water, says Douglass and others . Read his book for 271 humor-filled, well-documented pages of revelations about milk (and other whole foods).

Let me summarize a recent report from the Spring 1990 *HealthCare Letter.*[82] I have heard many concerned parents (whose children regularly drink commercial milk) say that they felt that the milk was safe because it is routinely tested for antibiotics. Yes and no. The standard test readily detects antibiotics in the penicillin family but is virtually blind to other drugs in wide use. Many farmers are aware of the "blind spots" and simply switch to less detectable drugs. A Rutgers University study reported in the *Wall Street Journal* tested 50 samples from ten cities using a more sensitive test which detects drugs other than penicillin. They found residues of other drugs, including sulfamethazine, a sulfa drug that causes thyroid cancer in animals; sulfadimethazine, a drug of unknown safety; and chloramphenicol, an illegal drug which in minute quantities can cause aplastic anemia in as many as one in 30,000 people. They also found erythromycin and streptomycin.[82] What about other drugs known to be used in this industry - tranquilizers, hormones, diuretics, worming medicines and other antibiotics not detectable by the more sensitive test?

What about skim milk? Drinking it to loose weight? Ask farmers what they feed their pigs to fatten them. You guessed it! Skim milk. Dr. Wulzen fed guinea pigs skim milk and they died, after developing wrist stiffness named the "Wulzen calcium dystrophy syndrome." In humans, this syndrome is called degenerative (calcific) arthritis (calcification of joints), hardening of arteries, cataracts and calcification of the pineal gland (which we now know is important in light physiology and hormonal regulation, especially in women). Why does this calcification occur? Among other things, calcium needs a bit of fat and an enzyme to be assimilated. Skim milk has neither. The enzyme has been killed (by heat) and the fat has been removed.

What about goat's milk? Fine, if it too is raw. Many people who cannot tolerate cow milk do fine on goat milk, which has several advantages over cow milk. The fat in goat milk is only one fifth the size of cow fat particles. Goat milk curd is small and flocculent; in cow milk it is large and dense. Goat milk is the closest substitute to human milk and is more easily digested than bovine milk.

From *Committee for Nuclear Responsibility Newsletter*. Author Unknown.

CHAPTER 14: USE ENZYME NUTRITION

Enzyme Nutrition

"Enzymes are substances that make life possible. They are needed for every chemical reaction that takes place in the human body. No mineral, vitamin, or hormone can do any work without enzymes. Our bodies, all of our organs, tissues, and cells, are run by metabolic enzymes. They are the manual workers that build our body from proteins, carbohydrates, and fats, just as construction workers build our home. You may have all the raw materials with which to build, but without the workers (enzymes) you cannot even begin." (Dr. Edward Howell, ***Enzyme Nutrition, the Food Enzyme Concept***, 1985).

Enzymes are very special proteins. They have been called biological catalysts, for no biochemical reaction occurs without them. But this term does not emphasize their biological force, vitality or enzyme activity, a quality that is measurable and influenced by age, temperature and pH (acid-base balance) of the organism in which the enzyme resides. As we age, our enzyme activity decreases. Also, enzyme activity depends on just the right pH and temperature range. Enzymes are much more heat-sensitive (labile) than vitamins and are the first to be destroyed during cooking, pasteurization, canning, microwaving, or heating above 118 degrees fahrenheit. You can swallow pounds of vitamins and minerals but without enzymes, nothing works and you are wasting your money.

Over three thousand enzymes have been identified. Enzymes can be divided into classes according to function. **Plant (or food) enzymes** occur in raw foods and initiate the process of digestion in the upper (cardiac) portion of the stomach. **Digestive enzymes,** made by the body, continue the process of digestion in the duodenum. Food and digestive enzymes also serve other purposes essential in maintaining health and vitality. **Metabolic enzymes** run the body chemistry and are involved in all body processes including breathing, talking, movement, behavior, and maintaining the immune system. **Antioxidant enzymes,** a special type of metabolic enzymes convert free radicals (reactive toxic oxygen species) into harmless products (water and oxygen).

Plant (or food) enzymes

These include protease, amylase, lipase and cellulase. These enzymes initiate the digestion of food. Digestion is a process of breaking food down into smaller units and adding water to it - a chemical process called hydrolysis. Each enzyme is very specific and digests only one type of food substance. Protease digests protein by converting it into its component amino acids. Amylase digests carbohydrates into smaller units called disaccharides. We know these by their common names, sucrose (cane sugar), lactose (milk sugar) and maltose (malt sugar). Lipase digests fats (lipids or triglycerides) into smaller units called fatty acids, plus glycerol (a trihydric alcohol). Cellulase digests the fiber cellulose into smaller units which includes D-glucose, a monosaccharide. Of these four enzymes, cellulose is very important because our bodies do not make it. So we must chew well the foods containing it to release the cellulase so that we can digest the fiber. People who are cellulase deficient have severe digestive problems including malabsorption.

Food enzymes function very differently from similar (pancreatic) enzymes produced inside our bodies (in vivo) and have other critical functions in maintaining health and preventing disease. The man who discovered this was the late Edward Howell, M.D., who developed his food enzyme concept in the 1930's after observing that people who ate raw foods had less illness than people who ate largely cooked foods. He wrote two fascinating books on his research, ***Food Enzymes for Health and Longevity*** and ***Enzyme Nutrition***. More recently, Humbart Santillo wrote a book called ***Food Enzymes*** for the lay public.

Dr. Howell's colleague, Howard Loomis, D.C. has continued this research at **21st Century Nutrition** (A). During the last ten years, Dr. Loomis has worked with food enzymes developed by the **National Enzyme Company** (A) and has discovered how to diagnose, treat, correct and prevent enzyme deficiencies in people who eat cooked and processed food diets. Dr. Loomis has developed a 24-hour urinalysis test to determine enzyme deficiencies in people. This is not to be confused with the Reams urine test which is based upon a random sample and does not take into account the hourly fluctuations in urine pH nor the effects of digestion.

Licensed health professionals and their assistants can contact 21st Century Nutrition for more detailed information and can attend seminars on how to use food enzymes.

Who needs food enzymes?

Everybody needs food enzymes, whether you eat only cooked foods, only raw foods or a little bit of each. Below I will attempt to tell you why it is important to use supplemental food enzymes if you eat cooked foods, and extremely helpful, even if part of your diet includes raw foods. By the way, raw food eaters are still enzyme deficient because today's plants, grown in depleted, sick soil will give up their enzymes to the soil in an effort to replenish it.

The references for all of this information includes standard medical physiology texts, such as Guyton's ***Textbook of Medical Physiology,*** in addition to Howell's and Loomis' research.

First, we need food enzymes because nature never intended the human body to produce them. In fact, food enzymes (protease, lipase, amylase and cellulase) work in the stomach but pancreatic enzymes elaborated by the pancreas work only in the duodenum and only when the duodenum is at the optimum alkaline pH. There is no overlap of activities nor will your pancreas become lazy if you eat food enzymes (as in raw food or food enzyme supplements). Instead, your pancreas will be spared from having to do 100% of the digestion in the duodenum. (Whew!).

Food enzymes work within the pH range (acid/base balance) of the human body, namely 3.0 (acidic) to 9.0 (alkaline or basic). The pH range of our gastrointestinal tract is about 1.5 to 9.0. Enzymes can only work in a narrow temperature range for optimal activity. As nature would have it, plant enzymes are at their peak activity at 92 to 104 degrees F, well within the temperature of the human body. As it turns out, these plant enzymes digest a large percentage of the ingested food in the upper (cardiac) portion of the stomach in the 15 to 60 minutes following ingestion.

The amount of each food digested varies according to what was eaten and the method of testing. But if you eat raw foods containing their enzymes or take plant enzyme supplements, then up to 80% of the complex carbohydrates, 60% of the starch, 30% of the protein and 10% of the fats are digested within 15 to 60 minutes of ingestion. This is the first stage of digestion, sometimes called predigestion, a process which occurs very inefficiently in anyone who eats cooked foods.

To repeat what I said above, do not confuse the predigestion of foods in the cardiac portion of the stomach with the second stage of digestion of foods in the duodenum. This happens when food enzymes are reactivated by the optimum alkaline pH of the duodenum. The reactivated food enzymes are joined by pancreatic enzymes. **Pancreatic enzymes do not work in the stomach but in the duodenum and only when the duodenum is alkaline.**

One of the biggest misconceptions in nutrition is the belief that food enzymes don't work in the stomach because HCl is there and destroys the food enzymes! **Wrong on both counts.** One, HCl is not released until the second stage of digestion, which occurs 30 to 60 minutes after eating. Two, plant enzymes are not destroyed but only deactivated by HCl, which decreases the stomach pH to less than 3.0 during the second stage of digestion. In fact, plant enzymes can be recovered intact from the duodenum where they are reactivated when the pH increases to a more alkaline pH, which is required for the third stage of digestion. I say 'when' but I should say 'if', because the duodenum only becomes sufficiently alkaline to activate food and pancreatic enzymes if there is enough bicarbonate available in the blood to be released to the duodenum.

Hydrochloric acid (HCl) comes from hydrogen ions (H+) and chloride ions (Cl-) in the blood and goes into the stomach via the parietal cells in response to the stretching of the stomach wall. The larger the bulge (e.g. the more food) the greater the HCl production. If you have plant enzymes in the cardiac portion of the stomach, partial digestion of food occurs and the stomach doesn't stretch as much, hence less HCl is needed.

In the third stage of digestion, the duodenum continues the process of digestion activated by an alkaline pH and the resulting secretion of pancreatic enzymes, including protease, amylase and lipase. As with HCl, the alkalinity also comes from the blood, specifically from bicarbonate (HCO_3^-), produced from its parent compound, carbonic acid (H_2CO_3) in the exact amount required to achieve the optimum alkalinity for adequate duodenal digestion. Mind you, I am not describing the whole story, which includes other important substances such as hormones, and other digestive enzymes (about 22 in all). I just want you to know these important points and to answer very common questions I receive in my practice.

Now comes confession of two of the cardinal sins of the nutritionist. It is wrong to give betaine HCl and other acidifying agents (glutamic acid, etc.) to acidify people who don't produce enough HCl and, on the other side of the coin, to give calcium carbonate supplements to alkalyze people who are too acidic. (I used betaine HCl myself until I discovered this).

To understand this you must know that the blood (the extracellular fluid or the ECF) is maintained at a very narrow pH range of slightly alkaline, 7.35-7.45. When I say the blood is overly acidic, I mean at the lower end of this range, e.g. an acid stimulus is being applied to the ECF. Overly alkaline blood is at the higher end of the above range, e.g. an alkaline stimulus is being applied to the blood. Anything higher or lower than this range is a medical problem, not a nutritional one. Remember this, we are speaking in a nutritional setting, not a medical one.

Acidifying supplements such as betaine HCl "overly acidify" the blood. In other words, they create an acidifying stimulus to the blood. In order to maintain homeostasis, this excess acid is neutralized by bicarbonate (alkaline) in the blood so there is none left to alkalyze the duodenum. Remember, the digestive enzymes elaborated by the pancreas can only become active when the duodenum is at an alkaline pH. So you can make digestive enzymes but they will not work without adequate alkalinity. This can only be achieved through the digestion of food, not by swallowing acids or bases.

In addition to wiping out the bicarbonate (with the resulting inactivation of the pancreatic and food enzymes in the duodenum), the kidneys must work extra hard to remove the excess acidity from the blood. This gives a urine that is overly acidic, as seen on a 24-hour sample. When I see an overly acidic urine, I immediately enquire about betaine HCl supplements and about eating (digesting) too many acidic foods (protein, especially animal protein; free form amino acid supplements; protein drinks; sugar and fat) plus other conditions associated with pathology such as diabetes.

The other important thing for you to know about HCl is that **HCl does not digest protein**, it only converts an enzyme, pepsinogen, to its active form, called pepsin. Pepsin is a proteolytic enzyme which works in the lower stomach at a pH of 3.0 and lower to digest very hard-to-digest proteins. Another way of putting this is: HCl can acidify the blood but it does not stop at homeostasis, so the kidneys must work to get rid of the excess acid. Also excess acidity (HCl) is neutralized by bicarbonate in the blood. Whereas if you eat a balanced diet of whole foods and supply food enzymes, the body will correct itself and the blood will maintain homeostasis through the digestion of foods.

Alkalyzing supplements, such as calcium carbonate, have the opposite effect. They "overly alkalyze" or supply an alkaline stimulus to the blood. HCl then becomes depleted in an effort to neutralize the excess alkalinity. This has much more far reaching effects than just turning off protein digestion. What amazes me is that even on TV you can see ads telling us to eat calcium carbonate to prevent arthritis and osteoporosis when in fact, the opposite is true, because by overly alkalyzing the blood you are laying the foundation for arthritis in any of its forms (including osteoporosis and osteoarthritis), plus other unhealthy conditions. Why? An "overly alkaline blood" does not carry adequate calcium (which requires both protein and a certain amount of H+ ions). This lays the ground work for calcium metabolism problems (e.g. arthritis, arteriosclerosis, cataracts), immune system problems (since hormones come from protein), and problems from protein deficiency.

A major problem from protein deficiency is hypoglycemia, because 56% of the digested protein is converted to glucose upon demand as the body needs it. Think of all the hyperactive kids and tired adults suffering from hypoglycemia who take drugs and live on adrenal stimulants (coffee, white sugar, etc). They could correct this condition simply by increasing digestion of good foods via taking food enzymes and/or eating more raw foods.

Again, how do we optimize the pH of the blood? This is done via the digestion of protein, fats, carbohydrates and fiber from a diet high in raw foods and from supplementing the diet with plant enzymes. Supply the blood what it needs via digestion and it will optimize its own pH. It will get acidity from proteins and alkalinity from vegetables and carbohydrates. Another way of looking at it is to think of your blood as being king because it must maintain homeostasis within a relatively narrow range of values. If it doesn't get what it needs from your food (because maybe you aren't eating food?!?) then it gets what it needs from your cells.

On the other side of the coin, it will dump things it doesn't need , such as extra chloride, calcium, etc., into your cells. In fact, cells will sacrifice themselves in order to maintain the homeostasis of the blood.

Now you know why plant (food) enzymes are so important and why they are different from the digestive enzymes made by the human body. These food enzymes are also very different from other plant enzymes, specifically papain (papaya enzyme), bromelain (pineapple enzyme) and pancreatin, a popular pancreatic enzyme derived from animals. First, bromelain and papain, though somewhat active in the human body, have their peak activity at temperatures much higher than body temperature and for this reason would be more suitable for meat tenderizing and the tanning of leather.

Pancreatic enzymes (pancreatin), whether eaten as a supplement or produced by the pancreas only work at a narrow alkaline pH range (7.8-8.3) and are thus totally inactive in the stomach. Again, if there is inadequate activity in the duodenum due to inadequate secretion of bicarbonate (often facilitated by taking HCl supplements), these pancreatic enzymes are totally ineffective.

Now, I will summarize the effects of a diet deficient or absent in food enzymes. (P.S., all diets are deficient in food enzymes!)

Effects of Enzyme Deficiencies

1. Pancreatic hypertrophy (enlarged pancreas). The pancreas enlarges in both animals and people who eat cooked foods. This has been observed by comparing pancreatic weights in wild vs domestic animals and in people who eat raw vs cooked foods.

2. Increased white blood count (called digestive leukocytosis) is observed in people who eat cooked foods. This occurs a mere 30 minutes after the cooked food is ingested. The more processed the foods, the greater the increase in the white blood count. Canned foods and processed meats cause the greatest increase of all. And guess who eats the most of these? Babies and children. Raw food causes no increase in the white blood count.

What do you think happens to the immune system when it is continuously mobilized every time you eat cooked, processed, canned, microwaved or pasteurized foods? Do you think that this would produce a stress upon it? Do you think that this would decrease its efficacy? Why do you think that 50% of Americans die of heart disease, one out of three people get cancer and that cancer kills more children than any other disease up to the age of 14? This constant stress on the immune system aggravated by environmental stressors (radiation, air pollution, pesticides, heavy metals, etc.) leads to an exhausted immune system and the degenerative diseases observed today.

3. High urinary indican. Colon toxicity is a direct result of maldigestion or the inability to digest foods. Whatever we do not digest will cause some kind of condition. Why? Visualize protein rotting in your colon. Yuk! Rotting organic substances produce a virulent toxin called indican in the colon. The colon, not desirous of this toxin, kicks it back into the blood. The kidneys whose duty it is to cleanse the blood, takes out the indican and adds it to the urine where it is easily detected via a simple test.

Indican is extremely toxic and causes many symptoms. The higher the indican, the greater the intestinal toxemia or inflammation in the digestive tract. Symptoms associated with high indican levels include: body odor, halitosis, foul odor of stool/urine, intestinal inflammation, gastritis, ileosecal valve incompetence, bloating, heartburn, diarrhea, constipation, malassimilation, weight loss, tachycardia, fever, allergies, asthma, arthritis, cardiac arrhythmias, ear/ nose/ throat problems, especially sinusitis, eclampsia, many eye problems, thyroid disease (goiter), nervous system problems, epilepsy, schizophrenia, memory deficit, phobias, melancholy, delusions, nightmares, senility, low back pain, sciatica, dermatitis, breast pathology and even cancer.[83]

I am emphasizing digestion here but please know that this is only the first step of metabolism. The digested food must then be absorbed across the gut wall and finally, assimilated (used) in the many processes of metabolism or in running the thousands of complex electrochemical processes required in the living organism.

Sources of Plant Enzymes: The National Enzyme Company has learned how to manufacture plant enzymes. You can find them in health food stores under such labels as Rainbow Light. However the health food store brands are only potent enough to initiate digestion of food. They will not correct a deviated blood pH or have affects on acute conditions, such as bee stings, pneumonia, infections, herpes, asthma, etc. There is only one source of plant enzymes strong enough to do these things. They are manufactured by the National Enzyme Company and marketed by NESS (Nutritional Enzyme Support Systems) (*). Since these enzymes are quite potent (enough to correct a deviated blood pH via digestion and to alleviate a crisis condition), they can be obtained only from licensed health practitioners who determine which enzymes are needed via the 24-hour urinalysis developed by Dr. Loomis.

Again, do not confuse this test with common lab urinalysis of your fasting morning urine or of analysis of a random sample as in the Reams urinalysis. The Loomis urinalysis is designed to determine exactly what kind of NESS plant enzymes are needed to corrected deviated body chemistry and bring the blood back to homeostasis. A drug may kill the germ that sneaked in when you were off balance, but it will not correct the condition that caused the imbalance. Therefore you will be susceptible to a repeat infection or condition that occurred during the imbalance.

In other words, we have two choices. We can take drugs to control the body or we can allow the body to control itself through optimum nutrition, which includes not just eating whole, organic foods, but digesting, absorbing and assimilating them! So eat as many organic, **raw** foods as possible and take plant enzymes. If you are off balance and your blood is battling to maintain homeostasis (equilibrium), find a Ness practitioner who can tell you the proper enzymes needed to bring your body back into balance.

Notes on the safety of raw foods: Many people ask me what kinds of raw foods should they eat and which are safe. In general, fruits, vegetables and dairy products should be consumed raw as much as possible. However, most grains (rice, millet, rye, barley, etc.) and red beans must be cooked and cannot be consumed raw. Nuts and seeds must be soaked before eating or lightly roasted to destroy enzyme inhibitors which protect the nut or seed from sprouting.

On the other hand, I do not think it is safe to eat raw meat (fish, chicken, pork or beef) because of the high level of pathogenic organisms they contain (parasites, bacteria, viruses and the cancer microbe, progenitor cryptocydes). To illustrate, there is a deadly bacterium called Vibrio vulnificus which commonly breeds in molluscan shellfish (such as oysters) which is 100% fatal if it goes undiagnosed and septic shock occurs. Thorough cooking kills this bacterium. The practice of cooking shellfish only until they open instead of for five or six minutes is not safe.

Another pathogenic bacteria associated with raw seafood and other foods is Listeria monocytogenes. This is a host-specific bacterium that afflicts pregnant women and causes the death of the fetus. These food-borne diseases are on the increase because of the increasing consumption of raw sea foods and sushi. [84]

Reprinted by permission of Universal Press Syndicate

CHAPTER 15: EFFECTS OF FOOD ENZYMES ON THE IMMUNE SYSTEM

Let me briefly outline other things that food enzymes do besides digest foods. When present in adequate amounts, these four plant enzymes can be and are used by the body to maintain health and prevent disease. In the acute conditions described below, these food enzymes are taken on an empty stomach so that, instead of digesting food, they enter into the blood stream where they do the following kinds of work.

Protease digests protein. Not just food protein, but other organisms which are also composed of protein, such as bacteria, and parasites. Protease cleans the blood of toxic debris such as that produced from poison oak or ivy. Because of its ability to digest protein, you can liken the function of protease to that of the white blood cells, specifically the neutrophils, the major component of the white blood cells. Dr. Loomis says that among food enzymes, protease is the first line of defense in the immune system.

Protease also quenches certain kinds of inflammation, such as that requiring ice to control swelling; fevers; and has a special healing effect upon the eyes, ears and kidneys. By increasing protein digestion, protease has an acidifying effect on the blood and increases blood protein, allowing the blood to carry protein-bound calcium as well as ionized calcium (Ca^{++}). Since 56% of digested protein is converted to glucose upon demand, having adequate protein digestion prevents hypoglycemia. Finally, people who are overly alkaline not only have joint pain, they are often extremely anxious. Acidifying to normal will eliminate this anxiety, often to the point where tranquilizers are no longer needed.

I have seen children whose immune systems have been weakened from the use of antibiotics to combat chronic ear infections make remarkable recoveries including recovery of infection-induced hearing loss, simply by taking a potent enzyme combination containing protease (from NESS). I have seen sore throats (from bacterial infection) recover with a single dose of protease, combined with enzymes containing a food source of the vitamin C complex (not ascorbic acid).

Amylase digests carbohydrates. It also digests white blood cells (pus). Because of its ability to digest pus, Dr. Loomis calls amylase the second line of defense in the immune system. Let me give you an example. If you have an abscessed tooth and antibiotics do not heal it, it's because there is no infection but only an abscess (pus with no bacteria). If you had adequate amylase, the amylase would digest the pus and the abscess would disappear within 48 hours.

Amylase is used in combination with lipase to digest viruses, including all types of herpes; to remove the pus from inflamed areas; and to control the type of lymphatic swelling that requires heat. The combination of amylase plus lipase will also help heal skin eruptions of all kinds such as hives, contact dermatitis, psoriasis; control the effects of bee and insect stings; and aid in clearing the lungs of hard, encrusted mucous accumulations. As you might guess, amylase in combination with other enzymes is an extremely efficient way of controlling asthma and other lung conditions.

My son, who developed asthma following exposure to pesticides and cigarette smoke, has not had an asthma attack since I began giving him these enzymes. In these clinical cases that I have described, most people do not have adequate levels of food enzymes to produce the above described results. However, the NESS (*) line of food enzymes are potent enough to achieve the effects I am describing.

Lipase digests fat. This includes fats in food and viruses which are composed of cells surrounded by fat. Lipase will digest the cell wall of the virus leading to its destruction. Dr. Loomis calls lipase the third line of defense in the immune system.

What do you think eating a diet high in lipase (or taking a high lipase enzyme supplement) would do in a person who has high cholesterol? Dr. Loomis believes that the Eskimos who ate a lot of fat had no heart disease because it was **raw** (high lipase) fat, not because it contained essential fatty acids. Other things contribute to high cholesterol, including a hypofunctioning thyroid and the inability to digest protein. Here I am merely emphasizing the importance of lipase in controlling fat metabolism.

CHAPTER 16: ANTIOXIDANT ENZYMES BY BIOTEC FOODS OF HAWAII

Antioxidant Enzymes: Superoxide Dismutase, Catalase, Glutathione Peroxidase and Methionine Reductase.

These enzymes do not digest foods. They convert free radicals (toxic oxygen species) back into the original materials, oxygen and water, without which we could not live. About 20 years ago a man named Jim Fahey whose family has been in farming and plant genetics since the late 1600's observed that every spring his cattle would heal when they started eating the tiny sprouts growing on the farm which they preferred over other grasses. He investigated these sprouts and discovered that they contained the above antioxidant enzymes. Over the last twenty years, Fahey developed hybridized wheat sprouts containing concentrated amounts of these antioxidant enzymes protectively embedded in the food sprout. These sprouts are harvested after a few hours growth, when their antioxidant content is highest. They are then dried to preserve their enzyme content and are enterically coated. Each tablet is equivalent to eating a tray of sprouts twelve inches by three inches in size.

By the way, these sprouts do not contain gluten, so gluten sensitive people can take them. For the super sensitive person, I found a secret that will allow even you to take these enzymes. Just take your plant enzymes, described in Chapter 14, with these antioxidant enzymes and you will experience no reaction to them!

These four antioxidants are manufactured by **Biotec Foods of Hawaii** and are actually foods but are formulated as tablets. They are just the protection we need in our modern age of radiation, jet lag, air pollution and chronic degenerative diseases. In the following discussion I will talk about certain properties of each of these enzymes as separate chemical entities. This is not quite accurate, because enzymes are integral parts of the tapestry of living tissue. Their functions are complex and related to the biological environment from which they arise.

In his recent book, ***Antioxidant Adaption***, Dr. Stephen Levine calls antioxidant enzymes "*the first line of defense*" against free radical pathology. Let me illustrate. Did you know that all is not dead inside nuclear reactors? There is a **living** organism, a bacterium called **Radiodurans** that thrives **inside** operating nuclear reactors. Guess what's in these bacteria? You guessed it! The highest levels of SOD/CAT and GP (plus others) ever measured! No, you don't have to scrape these bacteria off the reactor walls, thanks to Biotec Foods! No, they have not scraped them off either. They have developed high potency antioxidant enzymes from hybrid wheat sprouts - the only company to do this so far. Other sources of SOD are available but no food containing these enzymes has yet been developed like the Biotec antioxidant enzymes.

Now comes some incomplete research (P.S., all research is incomplete!). In general, the whole subject of antioxidant enzymes is swirling in a sea of controversy and scientific misunderstanding. The major lack of knowledge concerns just how these enzymes are absorbed through the gut mucosal wall. Some scientists believe these enzymes are too large to be absorbed. So let me tell you what has been discovered and leave to later writings the answer on just how it happens.

We understand much more about how these enzymes work than how they get into the body. Dr. Peter Rothschild M.D., Ph.D., did extensive medical studies trying to find out what was going on (A). So far, here is what he found. There is an unidentified "active" component of these enzymes which is absorbed across the gut wall. But, more importantly, the effect of taking these high enzyme sprouts is an unexplained stimulation of endogenous (i.e. inside the body or in vivo) production of the antioxidant enzymes. He showed this by giving patients the enzymes and doing blood studies before, during and after enzyme supplementation. Even after the patients stopped taking the Biotec enzymes, endogenous enzyme activity continued to increase.

Dr. Rothschild believes that these enzymes are electromagnetic substances that work directly with the cell membranes. Cell membranes have antenna-like pseudopods and receptor sites which communicate with the body. Immune system cells receive messages on how to function from neurotransmitters. Free radicals interfere with the neurotransmitters' ability to communicate with the immune system, so messages don't get through, leading to disease.

In all autoimmune diseases there are immune complexes, which are complexes of antibodies and antigens. These immune complexes accumulate in the blood and in the soft tissues which cause inflammatory reactions and tissue lesions. The antioxidant enzymes depolymerize (break apart) circulating and endogenous (soft tissue) immune complexes, formed from antibodies and antigens, which are always present in the soft tissue during inflammatory reactions of autoimmune diseases.

The healing may take time (even up to a year), but medical research has shown decreasing levels of immune complexes as the patient gets relief from his condition, and remarkable healings of rheumatoid arthritis and other autoimmune diseases have been recorded with this therapy. Arthritic pain decreases and sometimes vanishes. People with free radical pathology take them and get better. Radiation sickness is alleviated. Jet lag disappears. My major problem is convincing the suffering patient to take the enzymes long enough to produce relief. [85]

Blood chemistries of patients supplemented with the antioxidant enzymes have shown modulating effects on cholesterol, triglycerides and immune function, up or down as needed for optimization. In recent tests by the Swedish government on people with radiation sickness from the Chernobyl accident, Biotec enzymes were the only ones which produced noticeable relief. This is anecdotal clinical observation.

How Do these Antioxidant Enzymes Work?

Remember, Biotec's three enzyme formulations, SOD/CAT, GP and MR, each contain all of the others and are named according to the enzyme in highest concentration. I will briefly discuss each of these enzymes. However, I believe that it is difficult if not impossible to ascertain which one you need and recommend taking the combination enzyme formula, which contains equal amounts of all four enzymes. This combination formula is called Cellguard (in health food stores) or AOX PLX (*).

We've already mentioned how lucky we oxygen breathing mammals are to have SOD/CAT, natural enemies of the superoxide anion radical, ($O_2^{\cdot-}$). We talked about how $O_2^{\cdot-}$ is formed from an innocent looking oxygen (O_2) plus and extra electron; and how it loves to wreck your cell membranes by crashing through them. This critter is also one culprit that damages the joints during arthritis attacks. It is the free radical which reacts with hydrogen peroxide (H_2O_2) to form hydroxyl radicals ($HO\cdot$), a reaction catalyzed by xanthine oxidase (XO). Also formed by radiation, $HO\cdot$ goes on to make your fats rancid (yuk) and like $O_2^{\cdot -}$, damages the joints of arthritics and causes the pain of gout (along with uric acid crystals). Actually, HO· doesn't limit itself to your joints. It will attack any organic substance it runs into. Durk Pearson and Sandy Shaw talk about this in their popular book, ***Life Extension.***

SOD and partner catalase to the rescue! It's important to know that SOD will not work without catalase. SOD first quenches $O_2^{\cdot -}$ by converting it to hydrogen peroxide (H_2O_2) which catalase then converts to water and oxygen. There are no toxins produced during this detoxification reaction and thus none of the discomfort produced during the usual healing crisis. Clinical evidence records positive benefits from SOD/CAT in the treatment of arthritis (up to 90% remission), radiation sickness, jet lag, multiple sclerosis and as an anti-aging supplement. SOD/CAT also diminishes the side effects of radiation and the nausea of chemotherapy.

Glutathione Peroxidase (GP) is another important antioxidant enzyme system. It contains glutathione, a tripeptide composed of cysteine, glutamic acid and glycine plus selenium (four seleniums per molecule). Dr. Stephen Levine reported that oral selenium increases the activity of intracellular GP.[86] I have talked to several researchers who claim that Glutathione Peroxidase quenches both the hydroxyl and peroxide free radicals, HO· and HOO·, respectively. Glutathione Peroxidase and its partner glutathione have several known functions. One is to catalyze the conversion of hydrogen peroxide to water and oxygen, a reaction also requiring vitamin E. The glutathione is oxidized to its disulfide form which is then reduced to its active form by another enzyme, glutathione reductase. Catalase also converts hydrogen peroxide to water and oxygen but does not require glutathione.

Glutathione peroxidase/glutathione is great for the liver and many scientists believe that they are involved in protecting the liver from alcohol-induced liver damage. One way may be the glutathione peroxidase catalyzed reduction of hydroperoxides (HOO· and derivatives) by glutathione to harmless products before they attack unsaturated lipids, in addition to the reactions mentioned above.[87]

Free radicals can also damage fibrous protein within the body. For example, collagen is a protein comprising 30% of the body's protein, serving as a major structural component of skin, ligaments, tendons, muscles, cartilage, bones and teeth. The free radical cross-linking of collagen is commonly known as wrinkling. Many people taking glutathione peroxidase report improved skin appearance including shrinkage of moles and disappearance of pigmented (lipofuscin) "age" or "liver" spots and stretch marks as well as more evenly textured skin with smaller pores. Glutathione peroxidase is also reported to be an excellent nutritional support in the collagen diseases (lupus erythematosis, scleroderma, etc.)

Biotec's third antioxidant enzyme, Methionine Reductase quenches the most reactive of the free radicals, hydroxyl radical, HO· , mentioned above. Methionine reductase is especially good to combat chemical poisoning from carbon monoxide, insecticides, air pollution, etc. It is also effective in reducing athletic recovery time. Since MR quenches HO·, formed from radiation, it may be even better than SOD/CAT for alleviating the effects of radiation sickness. That's why I believe that the combination form of these enzymes (Cellguard or AOX) is the best form to take.

These antioxidant enzymes have other goodies in them that you may not realize. One of these is Coenzyme Q10, also called coenzyme Q, CoQ or ubiquinone (because it's a quinone present in nearly every living cell). CoQ has many cardiovascular and anti-aging benefits which are described below. There is research underway to find out how much CoQ is present in each of the antioxidant enzymes. Current information indicates that glutathione peroxidase has more CoQ than the other two.

Thoughts on using AOX (Cellguard) as a radioprotective agent during chemotherapy or radiation treatment for cancer

Cancer is an anaerobic process which produces destructive free radicals such as superoxide anion. In university studies of 50 different cancers, researchers found that all the cancer cells had little or no manganese-SOD, the specific enzyme that protects the cell nucleus from superoxides. Contact Biotec of Hawaii for research papers.

All the antioxidant enzymes quench radiation-produced free radicals. Therefore, I believe that using these enzymes and other radioprotective foods may be the difference between survival and death of the patient during these very toxic treatments. I also believe that these antioxidant enzymes protect only the healthy non- cancerous cells, whereas the cancerous ones are not protected. Therefore, these enzymes should not interfere or

Mainly because of research mentioned above indicating that cancer cells do not contain antioxidant enzymes or any other substance which supports aerobic processes.

I have had some experience with cancer patients who have refused to protect themselves in any nutritional way from their chemotherapy or radiation therapy because they believed that their cancer cells would be less vulnerable to the treatment. I believe that this is very wrong thinking. Indeed, the patients I knew who believed this died. I appeal to the research community to record the positive benefits of using radioprotective foods such as these antioxidant enzymes during the toxic treatments for cancer.

Use of Biotec enzymes in cardiac surgery patients

I would like to mention the use of the Biotec enzymes as preoperative nutritional support in cardiac patients undergoing cardiovascular surgery by Dr. Henry J. Younes, D.O. a cardiovascular anesthesiologist at Huntsville Hospital in Huntsville, Alabama.[88] One of the major causes of patient morbidity during open heart (bypass) surgery is due to severe tissue inflammation and free radical damage, a consequence of the surgical procedure. Dr. Younes found that giving the enzymes to his patients just 24 hours prior to surgery greatly improved the surgical outcome and reduced patient morbidity. I wonder what the results would have been had these patients been taking antioxidant enzymes for a long time prior to surgery instead of merely 24 hours?

Sources of the antioxidant enzymes: Biotec Foods of Hawaii is the only manufacturer of these antioxidant enzymes. It is available in health food stores, from health professionals, and mail order merchants, under the labels, Biotec, Biomed and Food Doctor, respectively (A).

Small quantities of SOD are present in certain foods such as barley (wheat) grass (1200 units per teaspoon), glandular extracts from liver, blue-green algae and yeasts.

New Product by Biotec Foods

Biotec recently introduced a new enzyme combination called **Fiberzyme.** It contains approximately one third each of the antioxidant enzymes, the same hybridized wheat sprouts grown for a longer time to increase the chlorophyll content, plus psyllium.

Fiberzyme is 20% soluble and 80% insoluble fiber. This combines the antioxidant activity of **AOX** or **Cellguard** with the properties of fiber and of chlorophyll (which are discussed in the next chapter).

Another new product contains the antioxidant enzymes to which have been added small amounts of another very potent antioxidant substance called pycnogenol, an extract from pine needles and bark. This substance has anti-inflammatory effects which benefit the vascular system and other effects similar to but more potent than the flavanoid family. Pycnogenol is described in Chapter 21.

> *"Harm in the form of excess human cancer, occurs at all doses of ionizing radiation, down to the lowest conceivable dose and dose-rate."...*
>
> from Dr. John Gofman's
>
> ***Radiation-Induced Cancer From Low-Dose Exposure***

CHAPTER 17: THE GREEN FOODS - BARLEY GRASS, WHEAT GRASS, SUN CHLORELLA, AND BLUE GREEN ALGAE

These are highly concentrated raw green juices of the grasses indicated, plus a red algae in the case of Sun Chlorella, and a blue green algae in the last case. They come in powder, liquid or pill form. I prefer the powder because its easier to assimilate and travels well. Some people prefer Sun Chlorella to Barley Grass. Others just love the Blue Green Algae. They all have similar but not identical compositions. I have observed that some people can tolerate Barley Grass but not Sun Chlorella and vice versa. All are excellent sources of chlorophyll and other essential nutrients. There are pure chlorophyll supplements available if you don't like to eat grass or algae but the above items are whole foods, not extracts.

I recently read an article from the The University of California, Berkeley Wellness Letter.[89] In it there was a very negative review of barley grass. The author stated that although enzymes are vital, enzyme supplements are useless when taken by mouth because all the enzymes you need are manufactured by your cells. Further, he stated that all the enzymes that you swallow will be destroyed by the digestive tract. Finally he says that *"chlorophyll does nothing for humans except turn their tongues green."*

This author clearly does not recognize the value, function and use of food enzymes, which have been medically proven to survive stomach acidity only to be revived into action in the alkaline environment of the duodenum. Secondly, the author does not recognize the value of chlorophyll in human healing and health. Thirdly, the author states that you can't get nutrition from barley grass that is comparable to eating each isolated food source of each nutrient. Think how many carrots, peas, and other foods that you would have to eat to get nutrition equivalent to one teaspoon of barley grass. For example, one teaspoon of barley grass is equivalent in nutrition to one whole head of cabbage!

Chlorophyll is plant blood, which closely resembles human blood and is used to cleanse, detoxify, purify and heal many conditions. It retards bacterial growth, detoxifies heavy metals from the body, increases wound healing, detoxifies the liver and other organs, deodorizes the body, removes putrifactive bacteria from the colon, aids healing of 11 types of skin diseases, relieves ulcers, gastritis, pancreatitis and other inflammatory conditions, helps heal gum diseases, and inhibits radiation and the metabolic activation of many carcinogens. Many studies have reported the protective effects of chlorophyll on irradiated animals. Please refer to the books by Russell- Manning and Hagiwara (B).

Let's talk about barley and wheat grass as whole foods. Though their compositions may vary in amounts of nutrients, both have a similar nutrient profile. There is brown rice in barley grass powder which will cause a reaction in a small number of people.

Wheat grass is usually available as a very potent freshly squeezed liquid. Remember, we are talking about grasses, not grains. Grain sensitive people can usually tolerate wheat and/or barley grass. I have never met anyone who cannot tolerate one or the other, regardless of allergies. In fact, the allergic person may experience a healing, especially those with skin problems such as eczema, acne, and rashes. The following information applies to barley or wheat grass. Read more about them in Dr. Yoshidide Hagiwara's booklet on barley grass or Betsy Russell-Manning's booklet on wheat grass.

Barley (wheat) grass is a totally balanced food. It contains all of the nutrients required for life - vitamins, minerals, enzymes and other proteins (amino acids), essential fatty acids and chlorophyll. You can drink barley grass juice anytime but it's absorbed fastest on an empty stomach. Some people will first experience a mild discomfort (diarrhea, headaches, etc.) due to the detoxifying action of the barley grass. After this initial discomfort, it gives energy and a feeling of well-being and mental clarity - all properties of optimum nutrition.

Here is a list of conditions which have been relieved or cured with barley grass: inflammatory conditions (peptic ulcers, gastritis, pancreatitis, arthritis, gout, etc.), hypertension, hepatitis, anemia, skin diseases, obesity, diabetes, hypoglycemia, fatigue, insomnia, hemorrhoids, high cholesterol, heavy metal body burden, prostate trouble, and PMS. These results were reported in clinical studies over the last ten years. In some cases the active ingredient which produced the effect was isolated, and sometimes identified.

Why is barley grass so versatile in its effect on the human body? For one thing, barley grass contains nutrients for every cell and organ of the body. It contains well-balanced minerals - calcium, magnesium, manganese, molybdenum, germanium, zinc, copper, selenium and lithium. Of these it is especially rich in potassium and calcium. It has an abundance of vitamins but especially beta-carotene (provitamin A), vitamin C, the B vitamins, (especially B-1, B-2, B-3, B-6, B-12) and vitamin E. One teaspoon (2 grams) of barley grass powder contains the equivalent of two heads of lettuce or one head of cabbage. Compared to an equivalent amount of wheat grain, barley grass contains five times as much protein, 26 times the potassium, 25 times the calcium, four times the B-1 and 100 times the vitamin C.

The high mineral content makes barley grass an excellent alkaline food. This is important because some people eat too many acidic foods (meat, sugar, fat) and this places an acidic stimulus on the blood (it should be slightly alkaline).

Barley grass has thousands of living enzymes. We have already discussed how enzymes are nature's biological catalysts that initiate all the chemical transformations in the body.

In addition to chlorophyll, other anti-inflammatory substances have been isolated from barley grass. For example, a fraction labeled P4D1 contains a protein (molecular weight 35,000) which has a stronger anti-inflammatory action than the anti-inflammatory steroid, cortisone, and non-steroid, phenylbutazone drugs.[90] Yet, it has few, if any side effects and no toxicity. Further, this P4D1 substance stimulated cellular DNA repair without creating any negative effect on its structure or metabolism.[90]

In one experiment, damage to cellular DNA was induced by x-ray irradiation and by adding a potent carcinogen. When these cells were incubated under normal conditions, some of them slowly repaired and some of them died. When P4D1 was added to the culture, DNA repair was promoted. However, adding P4D1 to undamaged cells did not cause any abnormal stimulation or suppression.[90]

A lectin-like material was also isolated from barley grass which stimulates DNA synthesis significantly, especially in older cells and tissues, hence the claim for the rejuvenating qualities of barley grass. In other studies, a substance was isolated from barley grass which lowers serum cholesterol.[91] In addition, large amounts of mucopolysaccharides were found in barley grass. Mucopolysaccharides (MP's) are polysaccharides (complex sugars) containing amino acids combined with simple sugars and sometimes protein. They have many protective cardiovascular effects such as lowering blood fats, retarding arteriosclerotic and aging processes within arterial walls - and are anti-inflammatory (relieve arthritis).[92,93]

If you have never tried barley (wheat) grass, this information may seem far-fetched. Don't take my word for it. Try it yourself and see what happens. I have observed many unexpected results from using it myself and in my practice. Just to mention a few examples: complete remission of eczema in a four-year-old, complete disappearance of bumps on the thighs and backs of the arms (due to insufficient vitamin A and/or E) in a seven-year-old, complete remission of sinusitis in a female (after battling with it for 20 years), normalizing of menses and remission of PMS in a 20-year-old-female, and many more. These results occur over a period of weeks or months.

The normal daily amount of barley grass powder is one to six teaspoons (one teaspoon = two grams) mixed with water or juice. If you choose wheat grass juice, start with one ounce daily and gradually increase to four or six ounces.

I consider barley (wheat) grass an essential protective food for children and introduced it to my twins when they were four years old. Remember, since children are more vulnerable to radiation damage and absorb heavy metals more readily than adults, this is a great food for them and is also a neat way to increase their levels of vitamins and minerals. Don't worry if they say "yuk" the first time they try it. Just keep reintroducing it in a positive way and they will form a daily habit of drinking their green juice. Keep mixing different juices until you find one pleasing to them. Try apple, orange, peach, orange-mango, papaya and grapefruit juices as well as vegetable and carrot juices.

Notes on Chlorella [94,95]

If you don't like either barley or wheat grass, try chlorella, a green freshwater micro-algae which has even more chlorophyll than barley (wheat) grass plus 55-65% protein with 19 amino acids including all the essential ones. It's an excellent source of beta-carotene, vitamins B-1, B-2, B-3, B-6, B-12, pantothenic acid, folic acid, biotin, PABA, inositol and vitamin C.

It also contains the important minerals, iron, phosphorus, magnesium, calcium, zinc, potassium, sulphur, iodine and trace amounts of manganese, sodium and chlorine. Chlorella comes in a small pill form which may be more palatable to some people. Sun Chlorella Company has an exclusive cellulose cell wall disintegration process which has increased digestibility to 80%

(which is up to 40% higher than other kinds of chlorella). You can find Sun Chlorella in most health food stores. Other forms of chlorella similar to Sun Chlorella have recently emerged on the market.

Notes on Blue Green Algae

I am particular about the source of algae because not all algae live in particularly clean waters. That is why most algae are grown under controlled conditions. The Klamath blue green algae is very special because it is a wild algae harvested under optimum conditions from its natural habitat in the Klamath Lake. Even this source of algae is harvested via many different methods, some of which are probably better than others in terms of product purity. It is wise to really investigate your source, in other words.

Blue green algae is similar to the green grasses above except it has more **low molecular weight neuropeptides** that feed the brain. Indeed, this algae has been called "brain food" and so profound is its effect on the brain that studies done in Japan show improved learning among learning disabled children. As well, many people noticed an increased mentality along with increased ability to understand certain topics such as, for example, mathematics. You can read the details of the effects of blue green algae on brain function and learning abilities in the book by Peter Tompkins and Christopher Bird entitled ***Secrets of the Soil.***

Don't run out and eat a bottle of this the night before your next math test. These green foods are potent in detoxifying the body and sometimes there is an initial side effect such as increased mucous, rashes, diarrhea, nausea etc.

Here is a great anecdotal story to illustrate what I mean. I saw it in an excellent video presented by Daryl Coleman of Cell Tech, one of the harvesters of blue green algae (A). One person who had prolonged exposure to asbestos broke out in a full-body pimply rash after taking blue green algae. After investigation of his condition, it was discovered that the center of each sore contained asbestos fibers! So if you take this stuff and all hell breaks loose, it may be that you're on your way out of hell into health heaven!!!

CHAPTER 18: PROBIOTICS OR FRIENDLY BOWEL BACTERIA, AN ENZYME PRODUCING FOOD, AND FOOD HYGIENE

Friendly bowel bacteria are called probiotics or micro-flora, and they have many beneficial functions in the intestinal tract including enzyme formation. Among them are Lactobacillus acidophillus, Lactobacillus bifidus, and Streptococcus faecium.

These are not enzymes but they help the body produce enzymes such as protease, lipase, and lactase, B vitamins, and natural antibiotics which inhibit at least 27 types of pathogenic bacteria. They thus confer protection from food poisoning and "traveler's diarrhea" (tourista), aid healing of skin diseases such as acne, help reduce cholesterol, improve gastrointestinal health, reduce the occurrence of urinary and vaginal infections, and are definitely radioprotective.[96,97]

Lactobacillus acidophilus is most commonly associated with cultured milk products (yogurt, kefir, etc.) but it is possible to get these friendly bugs as a supplement. The most common forms of lactobacilli are listed above. Studies have shown that lactobacilli lower the activity of fecal carcinogens and have anti-tumor activity.[98]

In addition, lactobacilli protect against yeast (Candida albicans) overgrowth, participate in B-vitamin syntheses, reduce lactose (milk sugar) intolerance, lower blood cholesterol, and produce natural antibiotics which protect against common pathogens (Eschericia coli, Salmonella typhimurium, Listeria, Clostridium perfringens and Staphylococcus aureau and Campylobacter pylori).[96,99] There is a growing body of evidence that peptic ulcers are caused by or aggravated by a chronic Campylobacter infection. Since L. acidophilus inhibits Campylobacter growth, it may be helpful in the prevention and treatment of peptic ulcers.[99]

Most people think they have to eat yogurt or other cultured milk products to get lactobacilli. Not true. Lactobacillus is a "lactic-acid-loving" bacterium. It converts carbohydrates to lactic acid. For example, sauerkraut and other fermented (lactic acid) vegetables contain lactobacilli. Those who say that lactobacilli can only be cultured in dairy media are referring to the early 1970 papers of Dr. Khem Shahani, a pioneering researcher in probiotics. Since then,

Dr. Shahani discovered that lactic acid bacteria, such as L. acidophilus, can be cultured in vegetable media. Vegetable-based probiotics are now available.

Let me describe S. faecium, a normal inhabitant of the intestinal tract, whose functions have been studied since 1965. There is an especially good strain of S. faecium which has been used in experimentation by many researchers including those in Sweden, Switzerland and Japan. The research has substantiated some highly desirable properties of S. faecium which include: fast growth, higher than usual resistance to acidity, quick production of lactic acid, good inhibition of harmful microorganisms, non-toxicity for animals and humans, and excellent effectiveness against diarrhea and intestinal infections.

Do not mistake S. faecium for S. faecalis, which produces hemolysis and toxins via the decarboxylation of amino acids.[100] Also, it is important to obtain the right strain of S. faecium, which is why I have mentioned Klaire Laboratories as an excellent source (see page 98).

Just a word about L. bifidus which is found in abundance in the healthy gastrointestinal tracts of nursing infants. This has been found, along with L. acidophilus, to inhibit the growth of common pathogens including coliforms and Salmonella. Along with L. Acidophilus, L. bifidus can improve the digestion of food and the biosynthesis of the B vitamins.

I take probiotics at least several times a week for all of the above reasons but especially to protect my family and myself from food poisoning which is now rampant world-wide due to terrible food hygiene. In fact, Dr. Leon Chaitow wrote an article on this subject, entitled "Food Hygiene".[101] In it he describes the growth of virulent (antibiotic resistant) bacteria from unsanitary food manufacturing processes. For example, antibiotics are used as feed additives to make animals grow faster and protect them from diseases resulting from packing animals tightly together in crowded quarters. In other words, antibiotics are used to promote growth and inhumane animal farming, not to treat disease.

By 1983, American livestock were eating nine million pounds of antibiotics. Today, about 50% of the antibiotics manufactured in the U.S. are fed to farm animals. These farm animals are over-confined, fed for rapid growth, given antibiotics, and allowed to wallow in feces. This leads to terrible disease epidemics.

What is the effect of the massive overuse of antibiotics? The creation of drug-resistant organisms which render penicillin and other drugs useless in the treatment of disease. It is a genetic law: antibiotics kill susceptible bacteria and the resistant ones survive, to pass on their drug-resistance to their offspring. As long as we use antibiotics, resistant microbes have a powerful advantage over weak competitors and can spread rapidly. These resistant organisms find their way from animals to people.

As these organisms become more virulent, they cause more damage. For example, Listeria monocytogenes is now 30% fatal. In his article, Dr. Chaitow discusses how it was found in coleslaw and 2% pasteurized milk. No one knows how Listeria survived the pasteurization process. Salmonella was also isolated from milk and eggs. Salmonella used to be outside the eggs. This was eliminated by proper washing procedures. Now it's inside the eggs, a direct result of our cultivating antibiotic resistant pathogens from over-using antibiotics.

The presence of salmonella in raw chicken is much greater than previously believed or reported. The U.S. Department of Agriculture recently reported salmonella bacteria in 57.5% of fresh poultry obtained from five processing plants in Georgia. Eleven of twenty one chickens purchased by the Los Angeles Times from five Los Angeles County and six Orange County markets tested positive for salmonella bacteria.

According to recent federal health statistics, salmonella is the leading cause of diagnosed food-borne illness in the United States. Two to four million people get salmonellosis each year and about 2,000 people die from it.[102]

Dr. Chaitow says that 8,000 people die each year from food poisoning, the direct result of modern agribusiness whose practice promotes the growth of salmonella and other pathogenic bacteria. Do you know that commercial feeds contain animal feces, powdered animal heads, feet and entrails? This practice repetitively reintroduces micro-organisms into the digestive tracts of animals feeding on such foods. Unnatural feeding and handling practices lead to epidemics of disease.

Let me briefly give you two graphic examples.[101] Bovine Spongiform encephalophy in British cows leads to madness and complete loss of nervous system control. It is caused by a virus which is inadvertently transferred to cows from sheep. How does this happen? It happens to cows who eat powdered sheep remains (forget that cows are vegetarians) to speed growth and increase their milk supply.

Ever wonder why so many commercial (not organically raised) chickens have salmonella? If you have a weak stomach, don't read this. After sacrifice, these chickens are plucked by rubber fingers to beat their feathers off. These fingers become smeared with chicken feces which are beaten into the pores of the chicken. The now filthy but de-feathered chickens are thrown into a vat where they float in a soup of feces, entrails and other dead birds plus an astronomical amount of bacterial contamination. The result? At least 35% and up to 70% contamination of commercial chickens with salmonella.[101]

Source of probiotics: Klaire Laboratories (*) has an excellent product called Vital-Plex (Prime Life in a health food store) containing three friendly bugs, L. acidophilus, L. bifidus and Streptococcus faecium, cultured in a milk-free (vegetable broth) medium. It contains over 10 billion live bacteria per gram (l/4 teaspoon) and is free of MSG (commonly used in probiotics), soy, grains, and yeast.

I am specifying Klaire Labs as the source of S. faecium because I have used it for eight years on hundreds of people including babies, pregnant and nursing mothers, children (including my own) and adults. I have read some papers condemning S. Faecium as a pathogenic bacteria. I believe that those who do this are either referring to a pathogenic strain of this bacteria or are confusing S. faecium with S. faecalis, a known pathogenic strain. In eight years I have never observed a harmful side-effect from Klaire probiotics but indeed have observed some positive effects, especially in very sick people, including young children who have been immune suppressed for years and have conditions including chronic diarrhea and other severe colon problems, severe yeast infections, chronic constipation, acne, digestive problems, parasites and other conditions. Information on probiotics may be obtained from Dr. Khem Shahani (A) and Klaire Laboratories (A).

CHAPTER 19: OTHER PROTECTIVE FOODS

1. Miso: naturally fermented soybeans, rice or barley. (Tamari is natural soy sauce from the liquid that rises to the surface of soy miso.) Miso is a superior source of usable whole protein, aids digestion and assimilation, is low in fats, has a wonderful salty flavor, promotes health and helps neutralize environmental pollution including radiation. The reason for Miso's protectiveness is not really known. It was observed that the Japanese who ate miso everyday did not die from radiation-induced cancer after the bombs at Hiroshima and Nagasaki. The observed effect of miso as a cancerostatic agent was further confirmed in a recent study in Japan. The researchers headed by Professor Shuichi Kimura, examined some 20 elements in miso paste and found that most misos contain 0.5% anti-cancer substances known as ethylesters of fatty acids. Kimura said that the amount of anti-cancer substances in miso is enough to eradicate the toxicity of five micrograms of benzpyrene, a potent carcinogen found in charred beef. It is believed that the anti-cancer substances result from the miso-making process and the combination of fatty acids in the beans and some types of yeast. Thus, raw soybeans do not share these properties. Miso should not be heated but added to the soup or other food after it is cooked, to prevent destruction of the enzymes present.[103]

2. Fermented (lactic acid) vegetables and juices: sauerkraut, beets, carrots, green and red pepper, beet tops, swiss chard and celery are especially good for pickling. According to Dr. Johannes Kuhl, an expert on the therapeutic value of lactic acid fermented foods, they possess medicinal properties and should be used in the biological treatment of many conditions, including cancer, arthritis, multiple sclerosis, kidney and liver diseases, and digestive disorders.[104]

As with milk, raw cultured vegetables and raw sauerkraut are much healthier than the pasteurized (heat-treated) cultured vegetables. Raw fermented foods (sold refrigerated) are rich in enzymes and are a delicious source of lactobacillus and lactic acid. They are an excellent aid to digestion. Rejuvenative Foods in Santa Cruz, California has an excellent line of these raw cultured veggies (A). I'm sure there are other sources in other parts of the country.

3. Cultured (fermented) milk products, yogurt, kefir, buttermilk, unprocessed cheese: In Dr. Douglas' book on milk (B), he described longevity studies indicating that groups of people who use cultured milk products in their diet live longer. Not the fake chemical commercial yogurt with lots of sugar, potato starch, artificial colors and flavors, and a few half-dead bacteria, but **real** yogurt made from fresh whole, raw, unhomogenized milk with billions of living friendly bacteria, a minimum of honey or fructose, and perhaps some **real** fruit. You can get it from companies such as Alta Dena, Honey Hill Farms, and Brown Cow. You can get raw kefir and raw milk from Alta Dena in California or Mathis Dairy in Georgia. Many people who are milk (lactose) intolerant can handle cultured milk products, in which the lactose has been converted to lactic acid.

The friendly bacteria, or probiotics such as Lactobacillus acidophilus and other strains perform valuable duties in your colon. I have already described some of these duties in Chapter 18. Let's summarize them here.

* These bacteria colonize the colon and evict unfriendly, dangerous bacteria and other organisms from your colon. You have probably heard of one of these unfriendly yeast organisms that became famous - Candida albicans - whose rise to fame and power has paralleled the increasing use of antibiotics-for-everything in America. A healthy colon full of these friendly bacteria will not allow yeast overgrowth.

* These talented friendly bacteria *synthesize their own antibodies in your colon.* The antibodies destroy germs that cause disease and food spoilage. Thus, they protect against infections.

* Lactobacillus acidophilus (and other milk fermenting bacteria) have been shown by the Sloan Kettering Institute for Cancer Research to have "anti-tumor activity." Thus cultured milk and its products are definitely anti-cancer foods. What about radiation protection? On page 254 of Dr. Douglass' book, he says, "*Dr. Tomic-Karovic exposed guinea pigs to X-ray. He found that those receiving cultured milk did not have abnormalities in their offspring. Those not protected by the milk had birth defects. Would it not be wise for every expectant mother to add kefir or yogurt to her diet?*"

* Yogurt, kefir and other fermented milks contain an unidentified substance which lowers cholesterol. Yogurt is an excellent substitute for ice cream in milk shakes and smoothies.

Use it on potatoes instead of sour cream. Make your own salad dressings with yogurt, honey, vinegar and spices.

4. Cruciferous vegetables (cabbage, brussels sprouts, broccoli, turnips, cabbage, spinach, cauliflower and greens such as kale): This family of vegetables appears to have anti-cancer value beyond their vitamin C and A content. They are thought to be protective because of substances which they contain (indoles) which inhibit cancer cell growth in the breast and the colon.

In addition, cabbage and other cruciferous vegetables contain dithiolthiones, a non-toxic group of compounds which were shown to have antioxidant, anti-cancer and anti-radiation properties. Researchers at Johns Hopkins University showed that mice fed dithiolthiones had a significant elevation in biochemical reactions which inhibit carcinogens and other toxic substances. Dr. Bruce Halstead, M.D., who has been actively involved in this research, said that dithiolthiones have the remarkable ability to selectively protect normal tissues but not cancerous tissues.[105] Dr. Halstead is no ordinary doctor. He is one of the world's leading toxicologists and his World Life Research Institute is an international toxicological center with one of the largest libraries in the world.

Dr. Halstead was stripped of his license to practice medicine, fined $10,000 and sentenced to four years in prison for using nutritional therapy (such as the famous South American herb, Pau d'Arco) with certain cancer patients. The Consumer Health Legislation League seeks to repeal the California law that Dr. Halstead was charged with breaking. This law makes it a crime for a physician to treat cancer with anything other than surgery, radiation or chemotherapy. This is one of the reasons why much useful information on natural methods of prevention and healing is not as accessible as information on synthetic drugs and conventional methods of therapy. And why most clinics who use "non-toxic" therapies are located just south of the border. You can get a list of these clinics from the Cancer Control Society (A).

5. Vegetables and fruits containing beta-carotene (can be converted to vitamin A in the body): This list includes dark, leafy vegetables such as broccoli, spinach, kale, swiss chard, romaine, endive, chicory, escarole, watercress, collard, mustard and dandelion greens; dark yellow and orange vegetables such as carrots, sweet potatoes, yams, pumpkins, winter squash; and fruits like cantaloupe, apricots, peaches, papayas, and watermelon. More on vitamin A and its precursors in the next section on supplements. Just know that foods containing them are good!

6. High fiber foods - whole grains, fruits and vegetables: Epidemiological data suggests that a high fiber diet protects against large bowel cancer perhaps for several reasons. It dilutes bowel carcinogens, decreases colon transit time and changes the composition and metabolic activity of the fecal flora and certain carcinogenic substances in the colon.[106,107] That's why it's healthier to eat **whole** grains containing the bran and the fiber, and **whole** fruits and vegetables instead of their processed, partitioned counterparts.

7. Foods containing a natural vitamin A analog, abscisic acid as pea shoots, lima beans, potatoes, yams, sweet potatoes, asparagus, tomatoes, onions and spinach; fruits like mango, grapes, avocado, pears, oranges plus the white under the peel and pulps, apples plus seeds and peeling, strawberries; all unsprouted seeds, especially sunflower, sesame and pumpkin; all nuts, especially almonds and cashews; leafy green vegetables; carob and teas from fruit blossoms and leaves (peach flowers, strawberry leaves, cherry flowers, apple blossoms); all grasses, e.g. wheat and barley. Read more about abscisic acid foods in Dr. Virginia Livingston's book, ***Food Alive; The Conquest of Cancer*** and others (A, B).

The vitamin A analog, abscisic acid is found in large amounts in the grasses, seeds, nuts, legumes, mature leaves, and fruits listed above and in the apricot pit. The body makes abscisic acid in vivo from raw carrot juice and raw liver, probably not as tasty as it is helpful! Dr. Virginia Livingston-Wheeler, M.D., who has a clinic in San Diego has spent a lifetime studying the etiology of cancer and has formulated a microbial theory of cancer (A).

In the 1960's, Dr. Livingston discovered a microbe she named Progenitor Cryptocides (PC). PC can be found in the placenta of a pregnant woman and in the tumor cells of humans and animals with cancer. Hence, Dr. Livingston-Wheeler called PC "the microbe of life and death." In a healthy person, PC, via its conversion to a hormone called Human Choriogonadotropin (HCG), repairs tissues and returns to rest. In an unhealthy or immune-compromised person, HCG fuels the growth of cancer cells. HCG is present in all tumor cells and is required for their growth. Next, Dr. Livingston-Wheeler discovered that abscisic acid neutralizes HCG. Now you know why the above foods are so important.

Dr. Livingston-Wheeler developed an autogenous vaccine from PC derived from the patient's own tissues, blood and urine which she gives back to him/her. She uses vitamin C and cell respiratory agents (B-3, B-2, B-12), immune system stimulants and a strict natural foods, nearly raw vegetarian diet with no meat, poultry or milk (raw cream okay). Her dietary program forbids poultry because it is nearly universally infected with the Cryptocides. This may or may not include locally produced, non-commercial poultry and fertile eggs. For her cancer patients she has developed "vaccinated eggs" which anyone can order (A) .

I believe that locally produced, organically fed, non-drugged poultry and their eggs are far less likely to be infected with the cancer microbe than commercial poultry. Dr. Livingston-Wheeler advises cancer patients not to drink milk because it contains a hormone, prolactin, which stimulates breast cancer. The acidophilus in unpasteurized milk MAY degrade this hormone (unproven hypothesis). On the other hand, raw cream and unpasteurized butter contain abscisins and other important nutrients not present in pasteurized cream.

Early in 1990, for nebulous reasons, the FDA confiscated her vaccine, despite years of documented scientific research proving its effectiveness in inhibiting the growth of cancer cells. Several months later, on June 30, 1990, Dr. Virginia died. Her final contribution is her latest book, ***The Hidden Plague***, which describes the infected, contaminated foods prevalent today.

Among the thousands of lives she touched is my family. In 1984, a very close family friend received a phone call from the doctor after a routine check-up, was informed that he had prostate cancer which had metastasized (spread) into his bones, and that he had only six months to live. Further, the doctor told him that nothing could be done to save him. Not pleasant news to receive via phone at the end of your day! There was nothing to do but seek "alternative therapies," so my friend went to Dr. Virginia for treatment. After her treatment, which included the vaccine, she referred him to a Canadian doctor, Ferdnand LaBrie, who gave him a treatment specific for prostate cancer. Today, totally free of cancer, my previously skeptical friend, has a healthy respect for nutritional medicine, non-toxic therapies, and the healing powers of his own body!

8. Sesame seeds, sesame tahini or the extract. Standard Process (*) has a product extracted from sesame seed oil containing an unidentified substance called Cataplex T. This substance comes from three primary sources: termites, meal worms and sesame seeds. The first two sources were ruled out! The fraction containing the "T" is concentrated and put into perles. The primary effect of Complex T is to increase blood platelets. This is especially important in cases of leukemia in which the blood platelet count is depressed. Two Cataplex T perles are equivalent to about two tablespoons of sesame butter.

If you don't like pills, eat sesame seeds or tahini. In addition to the substance, "T", sesame seeds contain valuable essential fatty acids (EFA's) which are important workers in the immune system.

9. Sea vegetables and their products, kelp, dulse (seaweed), algin (sodium alginate), agar (a red algae): Sodium alginate, a non-nutritious extract from Pacific kelp, is used to chelate (bind or complex) and detoxify heavy metals, such as lead, mercury, cadmium, etc. Agar is used as a thickening agent instead of gelatin or corn starch. Both will protect from radiation effects and will reduce absorption of strontium 90 by 50-80%. This is referenced in Paavo Airola's ***Handbook of Natural Healing*** and many other sources. Kelp and dulse, excellent natural sources of iodine, will help protect against radioactive iodine, found mostly in milk.

When there is adequate organic iodine in the diet, radioiodine is not as readily absorbed by the thyroid or the ovaries. Kelp contains 150,000 mcg of iodine per 100 grams (3-1/2 ounces). The RDA of iodine is 150-200 mcg. Other high iodine foods include: seafood, beef liver, pineapple, eggs and whole wheat.

Eating small amounts of kelp and other foods containing iodine is a lot safer than drinking iodine or eating potassium iodide, which can be especially dangerous for pregnant women and can cause allergic reactions. Caution: I have seen cases of thyroid inhibition due to overdosing on kelp so please use it in moderation.

Inorganic iodine is toxic because of its tendency to combine with protein. That's how it destroys bacteria (a protein). When you put iodine into your mouth, it combines with the protein there and in your stomach or wherever it goes. This causes irritation or worse. Also, sudden large doses of iodine in humans with a normal thyroid may reduce the synthesis of thyroid hormone.

What happens if you get zapped with radioiodine and develop a cancerous thyroid tumor? According to Robert Mendelsohn, M.D., "*even if you have a malignant tumor there would still be no reason to have it surgically removed unless it was causing pain or difficulty in swallowing. Thyroid tumors are self-limiting and don't metastasize, nor do they interfere with thyroid function. We have evidence of this from autopsies of Hiroshima survivors, all of whom had thyroid cancer from the radiation of the blast but didn't even know it. When surgery is prescribed, one should remember that the incidence of nerve paralysis to the vocal cords is about 10% and there is a* significant incidence *of permanent tetany (arm and leg spasms due to faulty calcium metabolism) and hypothyroidism..."* [108]

10. Bee pollen: This is an excellent food containing all essential amino acids, vitamins A, D, E, K, C, bioflavanoids, B-complex (especially pantothenic acid and B-3), and 27 minerals. I mention it here because of a study by Dr. Peter Hernuss at the University of Vienna's Women's Clinic, involving 25 women with inoperative uterine cancer. All received radiotherapy but the 15 who took 20 grams of bee pollen three times/day tolerated the radiation much better than the ten who took no bee pollen. This work is described in Earl Mindell's ***Quick and Easy Guide to Better Health.***

Dr. Emil Chauvin of the French Academy of Science found bee pollen beneficial in the treatment of anemia because it increases red blood cells and hemoglobin; chronic prostatitis; constipation; flatulence and colon infections, especially diarrhea.

11. Special teas:

Pau d'Arco (also called Ipe roxo, La pacho, Taheebo, and Bowstick tea): an herbal tea from the inner bark of two trees which grow in the warmer part of South America (Brazil, Argentina, Paraguay, Bolivia, etc.). Many South American doctors use a therapeutic tea from this herb to relieve pain and treat many conditions including leukemia and other forms of cancer, infections including yeast and other fungal conditions, skin rashes, and many other ailments.[109]

Dr. Theodoro Meyer of the Universidad Nacional of Tucuman, a province of the Argentine Andes, studied taheebo's chemical composition and found a substance called xyloidin, an antibiotic capable of killing viruses. Dr. Prats Ruiz, M.D., of Concepcion, a city in Tucuman province, reported blood profiles of patients with leukemia before and after treatment with taheebo to document its efficacy in treating certain cancers. Clinical details are provided by Professor Carlos Hugo Burgstaller in his book on the medicinal flora of Paraguay and Argentina, ***La Vuelta a Los Vegetales.***

This tea was widely available in many health food stores until news of its anti-cancer claims became widespread. There is now an FDA effort to prohibit its sale in the United States. It has already been taken off the Canadian market to be "reclassified as an over-the-counter drug" because of the healing claims made for it. FDA approval of a drug for a specific use takes 8-12 years and costs about $56 million (1982 estimate).[110]

Saffron tea contains colchicine, a substance which has been used in the treatment of leukemia. It also lowers uric acid which is high in those who have a tendency towards gout and/or cardiovascular disease.

Thyme tea. Use one tablespoon of thyme per pint of hot water, steeped 20 minutes, cooled and strained, as a remedy after x-rays. I could not find any information explaining its action or proving its effectiveness.

Black tea (containing catechins and tannin) and honey is used for radiation protection in China. Black teas contain catechins which protect against the damaging effect of strontium 90. Although I have read that tannin is another radioprotective ingredient, Dr. Steven Schechter (B) warns against using large amounts of black teas and says that excessive amounts of tannin may be carcinogenic!

"THE FIRST ACCIDENT WAS A RADIATION LEAK. THE SECOND ONE WAS A BIT MORE SERIOUS!"

CHAPTER 20:
MOTHERS, PLEASE NURSE YOUR BABIES

There are many benefits to mother's milk besides being always available, always fresh and always sterile. Nursing provides protection for the baby that cannot be obtained from any substitute. Breast-fed babies have more radioprotection, less infections, less eczema, less wheezing, less tetany (hypocalcemia, or depressed calcium levels), lowered incidence of crib death, less hypernatrenia (increased sodium) and fewer behavioral disorders. The nursing baby has less access to allergenic materials and breast milk itself is less allergenic because it's low in beta-lactoglobulin, a substance to which 80% of infants are allergic. Also, breast milk is antimicrobial. It contains anti-infective factors and antibodies.

According to Dr. R. Chandra, internationally known immunologist, breastfed infants have less infections, especially ear infection. He says nursing infants are more protected because they can't be left alone, whereas bottle-fed infants get milk in their ears and cow milk is irritating, which leads to ear infections. I believe it's because mother's milk is raw (high enzyme) milk which is not only more readily digested and assimilated, but has many other protective qualities, some of which I have listed below.

Breast milk has "survival selectivity". For example, it contains tiny amounts of iron which is fully utilized by nursing babies, preventing iron-deficiency anemia, whereas cow milk is well-known to inhibit the absorption of iron, thus leading to the need for iron supplementation in bottle-fed babies.

Breast milk may protect infant hearts. In a recent survey, heart disease was rarer in infants breastfed for at least one month or more than in bottle-fed infants. Breast milk contains more cholesterol than cow milk or infant formula. Researchers think that these high levels of cholesterol may help infants develop enzyme systems that enable them to handle fats as they grow older. Breast milk is also richer in essential fatty acids (EFA's) than cow milk. In fact, breast milk is one of nature's highest sources of a very important essential fatty acid, gamma linolenic acid (GLA). EFA's in neonates are very important for brain development and help maintain and repair arteries.

Breast feeding may contribute to straighter teeth because nursing infants use their tongues differently than bottle-fed infants. Nursing requires a forward tongue thrust and infants must use mouth muscles more vigorously. Infants nursed for one year had 40% less mis-aligned teeth than infants bottle-fed for one year.

Nursing is also important for babies with a high risk to allergies. Even so, one out of five breast fed infants will develop conditions such as eczema, allergies and asthma. There are several ways you can decrease the risk for allergies. These include:

* Exclusive nursing for the first six months and prolonged nursing for two years;

* If you don't nurse, use a hypoallergenic formula such as raw goat milk, sometimes with some raw carrot juice added to it;

* Delaying food introduction up to one year and especially avoiding wheat, eggs, animal milk, meat and peanuts ,

* Decreasing your baby's exposure to dust and other environmental pollutants as much as possible; and

* Consider using food enzymes by NESS or Rainbow Light to prevent colic and allergies from maldigestion.

The radioprotective benefits of mother's milk are presented in Chapter 4. If you want to know more about nutrition in nursing, read Betty Kamen's books, ***Total Nutrition for Breast Feeding Mothers*** and others (B). Also, contact your local La Leche league for loving support and nutritional information. There are, as well, many magazines for parents which emphasize wholesome foods and the nutritious value of mother's milk.

> *Forget the cringing selves that you sometimes are and remember, instead, the magic essence of your own being that sings even now through your fingertips. This is the reality you are seeking. Experience it fully.*
>
> Jane Roberts

CHAPTER 21: PROTECTIVE SUPPLEMENTS

As mentioned earlier, free radical pathology is one of the basic processes that occurs in most degenerative diseases. Free radicals and their products damage cells and enzymes, alter mineral concentrations in the cells, and decrease immune system competence. The immune system and the body's ability to defend itself against free radical attack depends on a good supply of antioxidants. Antioxidants are vitamins (C, E, A, beta-carotene, etc.); minerals (selenium, zinc, etc.); and other substances (enzymes such as superoxide dismutase or SOD, catalase, etc) which act as free radical scavengers. They trap free radicals and stop the chain reaction which otherwise would continue and cause cellular damage leading to mutations, cancer, etc.

Antioxidants have an active (reduced) form and an inactive (oxidized) form. For example the antioxidant form of the vitamin C complex is ascorbic acid. When it neutralizes a free radical, it is oxidized during the process to dehydroascorbic acid (DHA). Ascorbic acid is continuously oxidized to DHA and reduced back to ascorbic acid in the antioxidant recycling process. Other antioxidants, such as glutathione and bioflavanoids, are required in this recycling process to reactivate vitamins C, E and others. That's why vitamins C and E are much more efficient in doing their jobs if your diet includes selenium (as glutathione) and bioflavanoids.

Because antioxidants can also participate in the immune system, they are called immune system stimulants. For example, zinc is both an antioxidant and immune system stimulant.

A healthy body capable of withstanding radiation and other free radical sources must maintain adequate supplies of antioxidants and/or immune system stimulants. These nutrients are also required in everyday metabolism to build cells and provide energy.

These protective nutrients have a two-fold purpose: to decrease the incidence of carcinogenesis (transforming non-cancerous cells into cancer cells) and to increase immune system competence. Unfortunately, most people are overexposed to carcinogens and simultaneously have a depressed immune system. To repeat: antioxidants scavenge free radicals and stop the free radical chain reaction associated with cancer and other degenerative diseases. The immune system, if strong, will eliminate the cancer cells that have sneaked past the antioxidant/free radical scavenger system.

I have listed food sources containing each nutrient but I have deliberately deleted the dosage range. Why? Because I do not want to be responsible for the effects of taking the wrong dose of isolated nutrients for a prolonged period of time. This is between you and a competent health practitioner. I believe that any supplement or even any food used in high doses acts like a pharmacological substance, e.g. a drug, not a food. When indicated, such as during acute illness, this may be beneficial. But you need to know when this is indicated and for how long.

In other words, I believe that the pharmacological use of certain natural substances during illness may be appropriate because the body physiology changes at this time. To illustrate, people with cancer can tolerate an amount of selenium which would be lethal to a healthy person.

It is not wise to take megadoses of isolated nutrients, such as a single B vitamin, a single amino acid, high doses of ascorbic acid without the rest of the complex, high doses of vitamin E, etc. Why? **You should know that the nutrient that controls your body chemistry, is the one in lowest amount, not the one you are megadosing**. This is so for all nutritional substances. That's why I prefer food sources of vitamins and minerals combined with enzymes because only then can you be assured of getting full digestion and utilization out of what you are taking.

Taking megadoses of isolated nutrients can be a waste of money, because your body may not absorb them or have enough enzymes to utilize them. This is not just a thought in my mind. I have observed this in doing hundreds of 24-hour urinalyses according to Loomis in patients who are taking megadoses of calcium, ascorbic acid, their favorite B vitamin, etc. etc. Their urine shows low calcium, or low vitamin C, even though they are swallowing high doses of these supplements.

So why am I listing these supplements if I don't really think you should take them except under qualified conditions? Because I want you to know what they do in your body and how important it is to eat whole foods as much as possible so that your body may be nourished as nature intended.

If you are taking isolated nutrients, it is important to know the toxicity symptoms for certain vitamins, especially the fat soluble ones (A, E and D). Although water soluble vitamins have a lower toxicity than the fat soluble ones, if you take a lot of one B vitamin for example, without sufficient B-complex, it can cause side effects and/or an imbalance in the other B vitamins.

Minerals have a much narrower dose-response range than vitamins and are therefore potentially more toxic. For example, hyperferremia (too much iron) is becoming more common in our society, because (1) processed foods are loaded with it, (2) multiple vitamin/mineral supplements usually contain it, (3) people associate fatigue (which has multiple causes) with iron-responsive anemia and overdose on it, (4) people cook in iron pots and pans and (5) they drink water contaminated with excess iron and other minerals that are toxic in excess, such as copper, not to mention the toxic heavy metals.

So what's wrong with a little extra iron (or copper)? Iron and copper accumulate with age at sites where they catalyze free radical damage. Free iron promotes formation of the highly toxic hydroxyl radical which participates in lipid peroxidation (makes your fats rancid). Also, iron feeds bacteria and cancer cells and makes arthritis worse. That's why arthritics and other chronically ill people are anemic - nature is trying to help by taking iron out of the blood and storing it in the liver until you are healthy again.

Remember Dr. John Ott's observation of the clumping of red blood cells in people who use VDT's in rooms where there are no full spectrum lights or access to sunlight? Dr. Ott speculates that it is the iron in the RBC which makes this possible and questions whether VDT users should take iron supplements. A good question.

In my clinical work, iron supplements are not recommended unless serum ferritin levels are low. (Serum ferritin gives a reliable assessment of iron status). In fact I have seen sick people get sicker and develop infections after taking hematinics.

My comments on hyperferremia refer to hematinics and other sources of inorganic iron, **not** to organic sources of iron such as yellowdock, Pau d' Arco tea and echinacea. For example, NESS(*) combines very low doses of these herbs plus a small amount of organic (food-derived) germanium with food enzymes in a formula (**#17**) which has a profound effect in reversing iron-anemia within a matter of days as shown by blood chemistry tests. This type of iron would in no way be immune suppressive. I have seen this many times in my own practice with very sick children and adults alike.

The following list of antioxidants, free radical scavengers and immune system stimulants are important to optimum health and many of them work together synergistically. Some unique functions of each nutrient are listed as they pertain to radiation protection, antioxidant activity and the immune system. Again, I do not recommend that you take high doses of these substances. I am listing them here so that you will know what they do and in what foods to find them.

For more references, please read the National Research Council Report, ***Diet, Nutrition and Cancer*** (B).

Radioprotective supplements

1. **Glutathione:** This important free radical scavenger/antioxidant was partly described above as a substrate with glutathione peroxidase in the conversion of peroxides to harmless products. It has other functions as well. It protects against reactive oxygen species and other environmental toxins, including cigarette smoke,[111,112] peanut mold (aflatoxin) induced liver tumors in rats,[86] and radiation damage.[96]

Another property of glutathione is its ability to detoxify heavy metals such as lead, cadmium, mercury, etc. These are enzyme poisons and can cause severe damage to the brain, nervous system, immune system and organs.[113] Dr. Stephen Levine believes that its ability to quench free radicals makes it effective in the prevention of heart disease and cancer.[114] As an antioxidant, glutathione recycles vitamins C and E from their inactive (oxidized) states to their active (reduced) states and thus has a sparing effect on these vitamins.

In addition to these roles, glutathione has been used clinically in the treatment of arthritis and other inflammatory disorders, cataracts, diabetes, hypoglycemia, allergies, and eczema. Remember, glutathione is not a lone ranger. It requires the catalyst glutathione peroxidase (containing selenium) and the less talked about glutathione reductase (plus vitamin B-2 as its cofactor).

Sources: Glutathione is synthesized from three amino acids, glycine, cysteine and glutamic acid. The supplement, part of the Biotec antioxidant enzymes, is also available as a single nutrient health food stores.

2. **Coenzyme Q10 (Coenzyme Q, Ubiquinone)** is a naturally occurring substance similar to vitamin K and E in structure. It is found in the living cells of humans, plants and animals. CoQ is an essential component in the membranes of mitochondria, the "lungs" or "power plants" of the cell. Mitochondria are involved in

cell respiration and in the production of energy. It is just now becoming popular in the United States but has been used as a cardiac medicine in Japan since 1974 for about 6 million Japanese. CoQ is manufactured in the body but the amount made declines by as much as 80% as we age and this decline is thought to be linked to the aging mechanism.

CoQ is used for treatment of congestive heart failure, treatment of cardiac arrhythmias, lowering blood pressure, and reducing injury following stroke or heart attack. CoQ has anti-aging effects as well. It increases exercise tolerance, stimulates the immune system (which becomes less efficient with age), and is a potent antioxidant in its own right, inhibiting lipid peroxidation in mitochondrial membranes. It is also very effective in treating periodontal disease.[115, 116, 117]

Since CoQ10 has a profound affect on heart tissue, it is being studied clinically as a cardioprotective supplement during the administration of cardiotoxic drugs such as Adriamycin, used in cancer chemotherapy. [116,118]

Source: All living organisms contain Co Q in their cells. CoQ is present in the Biotec antioxidant enzymes. Health food stores also sell CoQ as a single nutrient.

3 . Amino acids are the building blocks of proteins and have other functions as well. It's important to have the complete balanced spectrum of amino acids required for protein synthesis and other special metabolic functions. Otherwise, things will get sluggish in the metabolic department.

There are at least three amino acids which act as antioxidants: cysteine, methionine and tyrosine.

As an antioxidant, cysteine, a sulfhydryl containing amino acid, works with thiamine and vitamin C. It is especially important in detoxifying acetaldehyde, a toxin produced form alcohol, cigarette smoke and smog. Note: don't just stuff yourself with cysteine. If you take cysteine and have inadequate vitamin C plus free iron or copper, cysteine may be oxidized to water insoluble cystine and lead to cystine stones in the kidney and urinary bladder. The hydroxyl radical is also produced in this reaction.[124]

Methionine, another sulfhydryl-amino acid can be converted to cysteine or taurine in the body.

Tyrosine is also an antioxidant amino acid and has several other metabolic functions. It is formed from the amino acid phenylalanine and serves as a precursor to melanin (skin pigment), adrenalin and thyroid hormones.[124]

Caution: I don't recommend taking free- form amino acids or single amino acids because you think you are low in protein or have some amino acid deficiency. Why? Because, amino acids and powdered protein supplements are acidifying, just like betaine HCl, and may cause an acid stimulus to the blood leading away from homeostasis. Also, as discussed above, the amino acid that controls the ball game (e.g. amino acid metabolism) is the one in the lowest amount, not the one you are megadosing. Instead of taking single amino acids, I recommend increasing your digestion of protein with digestive enzymes (Chapter 14) and such concentrated foods as barley or wheat grass which have naturally balanced amino acids.

I have observed two effects of taking high doses of either protein powders, free form amino acids or simply eating a high protein diet. The first is an asthma attack, even in people who never had a prior history of asthma. The second is a kidney toxicity problem indicated by an abnormal increase in the blood urea nitrogen resulting in kidney pain, even in young people with no prior history of kidney problems. In healthy people, these are reversible problems and the side effects disappear when the supplement is withdrawn. But in people who already have lung or kidney problems, this kind of supplementation can cause a more dangerous and even life threatening situation.

4. Essential fatty acids (EFA'S): Gamma linolenic acid (GLA) and Eicosapentaenoic acid (EPA): These work together, are destroyed in food processing and are important in the proper functioning of the immune system and in protection from cancer. GLA is formed from cis-linoleic acid in plant oils such as safflower, sunflower, and corn, etc. EPA is formed from alfa-linolenic acid in fish oils, cod and salmon especially. The conversion of cis-linoleic to GLA and gamma-linolenic to EPA requires the enzyme, delta-6-desaturase. This enzyme is inhibited (denatured, wiped out) by hydrogenated fats (margarine), high insulin levels (maturity onset diabetes), immune suppressive synthetic drugs (all of them are immunosuppressive), smoking, aging, alcohol, aspirin, epinephrine, radiation and carcinoid processes (cancer and other free radical reactions).

Why are EFA's important? They are starting materials for prostaglandins (PG's), hormone-like substances which do good things and bad things (depending of which one is around). For example, PGE1 (from GLA) is a vasodilator, inhibits thrombosis, controls blood pressure, regulates the immune system, is an anti-inflammatory agent, regulates brain function, and inhibits the proliferation of cancer cells. Key cofactors (minerals and vitamins) required in this conversion include zinc, magnesium, pyridoxine, niacin and vitamin C. Thus, deficiencies in these nutrients block delta-6-desaturase, GLA formation and ultimately PGE 1.[112, 119]

GLA may be one reason why vegetarian diets and the macrobiotic type diet are cancerostatic. Vegetarian diets contain cis-linoleic acid, the precursor to PGE1 via GLA. Animal food diets contain the fatty acid, arachidonic acid, the precursor to another prostaglandin, PGE2. PGE1 normalizes cell proliferation whereas PGE2 tends to proliferate cells. This selective alteration of fatty acid metabolism in vegetarian vs animal food diets may explain the anti-cancer effects of the vegetarian diet.[120]

EPA is also important because it blocks arachidonic acid, which forms inflammatory prostaglandins in addition to PGE2. Finally, EPA provides cardiovascular protection by decreasing total cholesterol, increasing HDL cholesterol, lowering triglycerides and decreasing platelet aggregation, the tendency for blood to clot.

One more thing. It's important that your food contain EFA's if you want to protect your skin from sun damage. Sunshine, vitamin D, EFA's and calcium all work together. To put it very simply, vitamin D helps transport calcium from your stomach into your bloodstream. EFA's help carry calcium from the blood into the tissues. Suppose you have vitamin D but not enough EFA's. You will develop "tissue calcium starvation." When this is severe enough, it causes itchy skin, intolerance to the sun, resulting in welts, and cold sores (herpes). Though herpes is a virus, its outbreak can usually be controlled by controlling calcium metabolism with sufficient calcium and other important minerals, vitamin D, EFA's and the proper glandular support.

Sources: Evening Primrose Oil, a natural source of GLA, is available for people who have trouble making their own GLA in vivo from cis-linoleic acid. Oops! Recently, the FDA took Evening Primrose Oil off the market because of what they considered to be "false claims." Not to worry though. Other sources of GLA are available, such as Borage oil, black currant seed oil, and other foods listed below. Max-EPA is available for those who need more EPA. Emulsified cod liver oil is also a good EPA source for those who need more vitamins A and D as well as EPA. Since GLA and EPA work together, it's good to take them both. See Dr. Richard

Passwater's books, ***Evening Primrose Oil*** and ***EPA-Marine Lipids*** for more information (B). AMNI (A) has a capsule containing both GLA and max-EPA called Balance +.

Food sources: There are various percentages of GLA and EPA precursors in plant, fish and bean oils. Warm weather plant oils contain a higher percentage of GLA precursors than fish oils, for example, safflower (70%), sunflower (60%), and corn (60%). Fish and bean oils contain a higher percentage of EPA precursors, for example cod liver, salmon oils and bean oils all contain about 30%. EPO contains 73% cis-linoleic acid and 9% GLA. Flaxseed (linseed) oil has 50% EPA precursor and 20% GLA precursor. See Passwater's books for more details (A).

5. Fiber supplements: Fiber is the indigestible parts of the cell walls of vegetables, fruits and grains. There are five types: cellulose, hemicellulose, gums (such as guar), lignin and pectin.

Fiber rich diets are already accepted treatment for diverticular disease and constipation, protection against gall bladder disease, and prevention of diabetes, heart disease and colon cancer. Different fibers do different things. For example, cereal fiber is more effective in lowering cholesterol and blood sugar than fiber from fruits, vegetables and nuts. Pectin from fruits (especially apples and oranges) and sunflower seeds is used to bind and detoxify heavy metals, radioactive isotopes and other toxins from the gastrointestinal tract. Dr. Paavo Airola (B) claimed that only pectin was found effective in eliminating radioisotopes.

You should know that the fiber contained in whole organic foods is much more effective than isolated fiber supplements in preventing conditions such as constipation and other diseases associated with low fiber diets such as cancer and gallbladder disease. Why do people eat refined foods from which the fiber has been removed and then rush to the health food store to get fiber in a jar or pill to alleviate the side effects of a low fiber diet? (I don't know. Do you?) Because they do, there are many colon supplements containing pectin and other fibers are available for detoxification and treatment of diarrhea and/or constipation.

Sources: American Biosearch (*) has one containing 14 ingredients, called Colon 14 which contains no laxatives and in fact can be used to break the laxative habit. AMNI (A) has a similar one, plus another one, Detoxiplex, especially for binding heavy metals. Biotec's Fiberzyme, contains one-third each of the antioxidant enzymes, a more mature and therefore higher fiber and higher chlorophyll-containing sprout, plus psyllium.

NESS (*) has an enzyme formulation (**#12**) which is high in cellulase, the enzyme which digests cellulose to smaller units. This is required before raw foods are digested and digestive problems can arise in people who do not chew raw foods well enough to release this enzyme. Also, this enzyme is used in cases of acute food allergies and in the detoxification of heavy metals. It will help to increase the absorption of digested foods, a major problem in our society where there is little natural food available.

6. Vitamin A: an anti-infection vitamin which is especially helpful in respiratory infections and is required for healthy eyes, skin, and mucous membranes. Epidemiological evidence shows an inverse relationship between blood vitamin A (retinol, from animal sources) as well as beta-carotene (provitamin A, the plant precursor to vitamin A) and cancer.[121] It is especially important in prevention of epithelial (skin) cancers and cervical cancer[122] - a logical finding since vitamin A deficiency leads to abnormal differentiation of epithelial tissues. Animal studies indicate that vitamin A has a protective effect against carcinogenesis by most, but not all chemical carcinogens.

Food sources: Liver, fish, fish oils, eggs, butter.

7. Beta-carotene (provitamin A): precursor of vitamin A from plants is not stored in the liver and does not have the toxicity of vitamin A. However, diabetics and people with hypothyroidism have trouble converting beta-carotene to vitamin A and may need the latter. In addition to its role as a precursor to vitamin A, beta-carotene has a role of its own as a free radical scavenger of singlet oxygen. It works under low oxygen partial pressure, such as in the lungs, whereas vitamin A works better at higher partial pressures. Thus, both are needed but beta-carotene has a special protective role against lung cancer and this is probably why carrot eating smokers get less lung cancer than carrot hating smokers! Research by Eli Seifter, a professor of surgery at the Albert Einstein School of Medicine in New York City, and others have shown that animals given beta-carotene survive radiation better than animals without it.[123]

Food sources: Yellow, orange and green leafy vegetables and deep yellow fruits.

8. B-Complex: anti-stress vitamins. These are also very important to the immune system and antibody synthesis (white blood cell production). White blood cell (WBC) synthesis is markedly decreased when one or more B vitamins are deficient, especially B-5 (pantothenic acid), B-6 (pyridoxine), and folic acid. Vitamin B-5, more effective with vitamin C, is essential for proper adrenal function.[124]

Food sources: Organ meats, milk and its products, fish, eggs, barley or wheat grass, brewer's yeast.

Notes on specific B vitamins in regard to radiation (free radical) protection, antioxidant and immune system functions

B-1 (Thiamine): B-1 the amino acid cysteine and vitamin C are used in an antioxidant system which protects against acetaldehyde, a free radical initiator formed in the liver from alcohol and found in cigarette smoke and smog.[124]

B-2 (Riboflavin): is required in making the enzyme glutathione reductase that recycles glutathione back to its active state.

B-5 (Pantothenic acid): Dr. Paavo Airola reported animal studies in which the survival rate following radiation was increased by 200% by giving pantothenic acid prior to exposure.[73]

B-6 (Pyridoxine): has been reported to relieve the nausea of radiation sickness.[124]

B-15 (Dimethyl glycine, also called pangamic acid): an antioxidant and immune system stimulant. Protects against toxic effects of chlorinated and fluorinated hydrocarbons and scavenges free radicals.

Inositol: protects against radiation damage . Dr. Paavo Airola (A) wrote that it binds DDT and reduces its harmful effects.

Para aminobenzoic acid (PABA): Many sources, including Airola, claim that PABA absorbs ultraviolet energy and blocks its mutagenic and carcinogenic effects.

Biotin: University of Pittsburgh researchers found that biotin-deficient animals developed a depressed immune response and a reduced thymus gland size.

9. Vitamin C complex: This subject is controversial and warrants its own chapter (see Chapter 22).

10. Bioflavanoids (Rutin, Hesperidin, Quercetin and others): are antioxidants required for vitamin C metabolism because they reduce the oxidized form of vitamin C back to its active form. These are potent anti-inflammatory substances used in treating asthma, athletic injury, ulcers, oral herpes, arthritis, etc. They reduce bruising from capillary fragility, decrease risk of diabetic cataracts, strengthen cell permeability and help immunize the body from various allergies, especially hay fever, poison oak and poison ivy. Since bioflavanoids strengthen cellular integrity, it is not surprising that they offer protection against radiation. Animal studies verify this. For example, rutin (from buckwheat) reduced death fro x-rays in animals by 800%. Books by Jeffrey Bland, Linda Clark and Paavo Airola have information on bioflavanoids (B).

Food sources: Lemons, grapefruit, oranges, tangerines, grapes, plums, cherries, blackberries, rose hips, black currants, peppers, buckwheat, plant leaves.

11. Vitamin E (Natural form, d-alpha tocopherol, succinate or acetate derivatives; synthetic form, dl-mixtures): an antioxidant and natural chelating agent which aids in the absorption of vitamin A, along with zinc. Vitamin E participates in the glutathione peroxidase antioxidant enzyme system. It protects an important anti-inflammatory prostaglandin (PG12) from destruction by free radicals (fatty acid peroxides). PG12 is important in preventing thrombosis. Like beta-carotene, vitamin E traps singlet oxygen, however, it does so at higher concentrations than beta-carotene, so both are needed. Vitamin E was recently linked, along with beta carotene, as a protective agent against lung cancer. At Louisiana State University, Lopez and LeGardeur found that lung cancer patients got 50% less vitamin E from their diets than the healthy control group.[125]

Vitamin E's protective role against cancer may be related to its ability to preserve the integrity of red blood cells; its effect on connective tissue, which reduces contracting and tenderness of scar tissue during healing; its ability to reduce the oxygen requirement of tissues by increasing survival of cells during oxygen deprivation; and its role in preventing fats from becoming rancid.

Note: people with rheumatic hearts or high blood pressure should increase vitamin E dosage gradually because of its effect on circulation which could cause a sudden increase in blood pressure. Information on the function of this and other vitamins and minerals may be found in the seminar tapes of Dr. Jeffrey Bland and the seminars given by Drs. Jonathan Wright and Alan Gaby (A).

Food sources: Cold pressed oils, wheat germ/ oil, sunflower seeds/oil, whole grains, butter (especially raw), green leaves. This is my first choice, rather than taking any dose of an isolated form of the vitamin itself.

12. Zinc: an antioxidant mineral and immune system stimulant, required for proper immune system function, growth and healing. Zinc is part of the superoxide dismutase antioxidant enzyme system (Chapter 16) which quenches free radicals. It participates in over 100 enzyme systems. During healing, zinc chelates dead viruses, bacteria and other cellular debris and carries them out of your body. It mobilizes vitamins from liver storage. As part of the insulin molecule, zinc participates in carbohydrate metabolism as well as protein metabolism and male gonad function. Thymus hormones, which are involved in immune system functions, are altered in zinc deficient people, and their thymus weight decreases.[81]

Food sources: Oysters (highest food source), ginger root, lamb, liver, nuts, split peas, egg (yolks), whole grains, lecithin.

13. Selenium: The most important known function of selenium is its participation in the glutathione peroxidase antioxidant enzyme system. It also functions synergistically with vitamin E making combined effects greater than additive; and with cysteine and other sulfur-containing amino acids.

Epidemiological studies show a lower incidence of cancer in areas where the soil is high in selenium.[126] Selenium stimulates the immune system, helps maintain cell respiration, detoxifies mutagens, carcinogens and heavy metals (especially mercury), and protects from free radical (radiation) damage according to the National Research Council report on ***Diet, Nutrition and Cancer***. There are two forms: organic (selenomethionine, selenocystine) and inorganic (sodium selenite, selenate). However, most organic sources are reported to be less effective than selenite in elevating deficient levels of glutathione peroxidase. Selenized yeast is reported to be an exception to this. The inorganic form is used in cancer therapy and has a profound effect on tumors. Animals with cancer need many times the dose of selenium required in healthy animals. This effect goes beyond the amount of selenium required in normal (e.g. glutathione peroxidase) metabolism.[82] Research indicates that selenium-fortified kelp can reduce and sometimes eliminate chemical sensitivities to hydrocarbons, formaldehyde, chlorine and phenols, many of which act as free radicals.[114]

Caution: Normal, healthy people should not take high doses of selenium as a preventive measure against cancer.

Food sources: Butter (highest source), herring, smelt,

wheat germ, brazil nuts, apple cider vinegar, shellfish, whole wheat, bran.

14. Germanium: Germanium greatly enhances the effectiveness of oxygen in the body and Dr. Kazuhiko, a Japanese physician who studied Germanium for years, found it effective in the treatment of many degenerative diseases including cancer, diabetes, hypertension, liver cirrhosis, epilepsy and arthritis. This is related to the fact that cancer is an anaerobic process (requires no oxygen), and in fact, all chronic degenerative diseases are associated with a lack of available oxygen in the body cells.[127] Information on Germanium can also be found in Dr. Betty Kamen's book by the same name (B)

Caution: Great Britain has classed Germanium as a poisonous drug and has taken it off the market, claiming that long term use causes kidney problems.[130] Whether this has been proven true or not, it illustrates an example of the caution I am recommending in using high doses of isolated supplements for any length of time.

It is logical to me that any substance that is used in high doses can cause harm to any organ in the body but especially the organs of detoxification and elimination, namely such as the liver, kidneys, colon and skin. That's why so many drugs cause skin rashes, antibiotics lead to yeast infections from their over-kill effect on the intestines and most drugs that control body functions have a harmful effect on the liver and/or kidneys. When you take a high dose of even a nutritional substance for any length of time it is no longer a food. It is a pharmacological substance - i.e. acts like a drug, not a food.

Let me add my opinion on the toxic effects of Germanium and leave it up to science to see who's right! I believe that the toxic effects observed with Germanium are caused by high doses of inorganic Germanium, which I consider wrongful use of a healing substance. As evidence, I offer the following evidence from my experience as a NESS enzyme therapist. The NESS formula **#17**, which I described as an organic source of iron, also contains a small dose of organic Germanium. It is used not only in the treatment of iron-deficiency anemia, but also for migraine headaches, children's mild complaints (tummy aches, intermittent fevers), viral conditions and lung conditions, including asthma and emphysema. I have seen all of these conditions disappear with Ness **#17.** The remedy is used only as long as the particular condition exists.

Food sources: This rare earth mineral is found in ginseng, certain mushrooms, garlic, bamboo sprouts, and certain other plants.

15. Glandulars: The endocrine and exocrine glands secrete hormones and enzymes which are extremely important in regulating body metabolism. The thymus and spleen are involved in immune system functions but the other glands (thyroid, pancreas, adrenals, liver, pituitary, etc.) are also important.

Glandular therapy is used as an adjunct to vitamins, minerals and enzymes. For example, raw adrenal glandular provides t nutritional support for the hypofunctioning adrenal gland in addition to other adrenal foods (containing pantothenic acid, vitamin C and zinc). The concept, similar to homeopathy, is "like cells help like cells." Raw glandular tissues from animals contain unidentified intrinsic factors distinct from vitamins, minerals, enzymes, or hormones. These cellular factors are cross reactive (from animals to humans) and organ specific. Thus cellular factors from bovine thymus will be picked up by a human thymus. This has been demonstrated in studies by Dr. A. Kment in Germany through radioactive isotope studies.[128] Using glandulars improves the function of the particular gland by delivering gland-specific substances to it.[129]

I no longer use glandulars because of the effectiveness in using plant enzymes to increase the digestion, absorption and utilization of whole foods.

Caution: There are researchers in England who believe that glandulars derived from beef and sheep should not be taken due to the probability of their contamination with bovine spongiform encephalopathy (BSE) and other pathogens.[130,101] Heat treatment may (?) kill BSE but it also destroys the enzymes and changes the structure of the protein, making therapeutic effect less doubtful. Contact Simon Martin of *Here's Health* (A), and Leon Chaitow (A,B) for more details. There are fears that the U.S. now has BSE in its cattle, despite a ban on imports on infected cattle from Britain.

16. Raw bone marrow (Calcifood, from Standard Process (*): Standard Process is the only manufacturer of **raw** bone marrow that I can tell you about. Calcifood is uncooked whole bone marrow with all the marrow still in it, totally unlike steamed bone meal usually sold elsewhere. Steamed bone meal is what's left over after boiling the bones and selling the dried juice as gelatin. Everything that is not water soluble is left in the bone solids which are dried, ground up and sold as veal bone. Calcifood is an excellent nutritional support for the bone marrow. It helps build red blood cells and white blood cells.

The thymus and spleen are also active in the formation of other types of white blood cells. People who have a low red blood cell count, or faulty production of white blood cells, or who need a balanced source of calcium and phosphorus for bones and teeth may benefit from Calcifood. Strong healthy bone marrow is more resistant to radiation (free radical) damage than weak unhealthy bone marrow. Of course, the caution above regarding BSE contamination of beef derived products applies here as well.

17. Homeopathics, classical (traditional) and complex: Are there homeopathic remedies (other than the traditional constitutional remedies which are individually prescribed) which aid in protection from radiation damage? The data on this are scarce. There are literature reports of homeopathic remedies which successfully protected children and adults who were exposed to scarlet fever, whooping cough and polio. For example, a homeopathic preparation of whooping cough bacillus given to 364 exposed children was 100% effective in preventing the disease.[145] See Dr. Randall Neustaedter's new book, ***The Immunization Decision*** (B), for the latest details on this subject.

There are clinical reports of people desensitizing themselves from allergens in foods or the environment, such as poison oak or ivy with homeopathic preparations of the particular allergen. Marie Curie's father, a classical homeopathic physician, believed that Rhus vennanatta (30x) taken over a period of several weeks would counteract the effects of radiation exposure. This is a controversial area. Some classical homeopaths claim that constitutional treatment to enhance the immune system combined with homeopathic preparations during direct exposure to disease or radiation is the only homeopathic method which works.

On the other hand, other health practitioners believe that complex homeopathics (the use of certain remedies for certain conditions) is also effective either alone or combined with the classical remedy. For example, a person with candidiasis (yeast infection) would be administered homeopathic Candida albicans. Complex homeopathics is a relatively new addition to a time-honored field. The only written information I could find was by Julian Kenyon, M.D., N.D., who believes that complex mixtures have two effects: they minimize side-effects that may occur by giving a single remedy, and they may give a wider range of action than a single remedy may give on its own.

In his book, ***Modern Techniques of Acupuncture,*** Dr. Kenyon postulates that homeopathic nosode complexes release toxins from the cells into the extracellular fluid (e.g. outside the cell) from where they must be excreted via the body's excretory mechanisms. Sometimes a cleansing homeopathic or herbal remedy is combined with the nosode to facilitate elimination of the toxin. Complex homeopathy is more popular in Europe than the United States and there are more complex homeopathics available from there.

Currently there are several companies who manufacture homeopathic remedies to protect from or detoxify the effects of radiation. Do these remedies actually provide radiation protection? No one can say either 'yes' or 'no' and prove it. Currently, practitioners try to determine whether a person needs the remedy via kinesiology (a sophisticated type of muscle testing practiced by chiropractors) or other methods such as a Voll-type EAV machine.

The common belief that homeopathic "immunization" is only effective during an epidemic (i.e., during actual exposure to the disease) is irrelevant in this context. ***Radiation is an epidemic. There is no such thing as an unexposed person.*** The only questions are how much and how protracted is the dosage and from what source?

As you can see from the vague explanation above, more documented research needs to be done (and fast!).

18. Dehydroepiandosterone (DHEA): DHEA is second only to cholesterol in abundance in the human body. It is being studied as a possible life extension food extract and studies have shown that oral DHEA prevents obesity and cancer in laboratory animals and extends their life span by 50%. The theory is that DHEA inhibits cancer through inhibiting enzymes that transform potential carcinogens into active carcinogens.[131] Thought you'd like to know about this in case it shows up in your health food store.

Food source: Yams.

19. Pycnogenol: a patented substance extracted from pine needles and bark which contains 85% proanthocyanidines by weight. These are believed to be the "active" antioxidant/free radical scavenger substance in the extract. These substances are supposed to be similar in structure to flavanoids but their activity exceeds that of the flavanoids in similar areas of activity. For example, both increase capillary resistance and enhance the function of vitamin C. However, pycnogenol appears to have more than an enhancing effect. Studies have shown that pycnogenol is 50 times more effective than vitamin E and 20 times more effective than vitamin C as an antioxidant in addition to having the ability to potentiate the antiscorbutic effect of vitamin C. Pycnogenol's ability to increase capillary resistance makes it important in the prevention of vascular conditions such as: varicose veins, diabetic retinopathy, peripheral hemorrhage, high blood pressure and heart disease.

The anti-inflammatory properties of pycnogenol make it useful in the treatment of hemorrhoids, swollen joints and sports-related injuries. In addition, pycnogenol has a role in the prevention of cirrhosis of the liver. [132,133]

Remember your attitude, thoughts and feelings. They have a profound effect on your immune system. Despair, fear and depression are just as unhealthy as junk food.

CHAPTER 22: THE VITAMIN C COMPLEX[134,135,136]

Many people believe that vitamin C complex and ascorbic acid are the same. They are not. Ascorbic acid is only one part of the vitamin C complex. In nature, ascorbic acid is the antioxidant part of the vitamin C complex and as such is bound to a protein called vitamin P, of which Rutin is a part. The **vitamin P factor** of the vitamin C complex is involved in preventing hemorrhage and capillary fragility but there are four other factors as well. One is called the **K factor** which aids in blood coagulation. The **bioflavanoid factor** improves capillary strength, coagulation of blood and the absorption of nutrients across the gut wall. The **J factor** carries oxygen and activates cellular ferritin. In the **ascorbagen factor** is tyrosinase, a copper-containing enzyme involved in several reactions with tyrosine, an amino acid precursor of epinephrine, norepinephrine (adrenal hormones), melanin (skin pigment) and thyroid hormones. Picture these five factors as spokes on a wheel and ascorbic acid as the tire on the wheel. Ascorbic acid is the antioxidant that protects the other factors from being oxidized. The other factors act as oxidizing agents. This oxidizing part of the vitamin C complex is involved in the electron transport chain in which proteins are desaturated to form double bonds via the transport of electrons, and then oxidized to produce **energy**.

As you can see, taking only ascorbic acid without the other parts of the complex is just like taking any other isolated nutrient. It causes the body to "pull the spokes off the wheel" in order to obtain the other parts of the vitamin C complex. This leads to a deficiency of these elements in the body. People who can tolerate high doses of ascorbic acid are those who have high amounts of the other spokes on the wheel in their bodies. Although vitamin C complex is not stored in the body like other vitamins, it is found in large amounts in leukocytes (white blood cells), collagen, adrenal and spleen tissues. In particular, collagen is a protein containing 60% vitamin C complex. Since collagen comprises 30% of the protein of the body, it can supply ascorbic acid with the other parts of the complex. This, in time, will cause a collagen deficiency and other deficiency symptoms as well, especially in those who use megadoses of ascorbic acid on a daily basis.

The large amount of vitamin C complex in collagen tissue is one of the reasons why vitamin C is known as a protector of connective and endothelial tissue. Vitamin C aids in weaving cells more tightly together and activates the constructive phase of the protein-building enzymes, explaining why vitamin C deficiency causes marked deterioration of the mucous membranes and gum tissues of the mouth and why wound healing accelerates in people who have adequate vitamin C. Vitamin C also influences the metabolism of calcium and phosphorus. So a deficiency of vitamin C and minerals in the diet will cause imperfectly formed teeth as well as gum disease.

The white blood cell (leukocyte) carries more vitamin C than any other cell of the body. Thus, vitamin C protects the leukocyte structure in its biochemical battle with invading microorganisms.

Vitamin C complex has many other functions which may be unfamiliar to some people. For example, vitamin C is required for optimum utilization of oxygen by the blood and is thus one of the requirements for cardiovascular health. Vitamin C complex also combats the toxic effects of hyperthyroidism which involves the deamination (removal of nitrogen) of proteins by thyroxin. Vitamin C prevents decomposition of the acid, adenyl pyrophosphoric acid, essential to heart muscle function. This acid decomposes with undue thyroid stimulus. This also explains why vitamin C complex combats shock. The deaminizing enzymes from the damaged cells, which release histamine and other shock producing end products, are inhibited by vitamin C. Vitamin C also reduces the need for high fever during infection because it stimulates phagocytosis which is also facilitated by heat. With the added help from vitamin C, the high temperature becomes unnecessary.

Systemic diseases and all infectious processes exhaust the reserve of vitamin C and often initiate rapid degeneration of dental structure or a repeated bout with the illness, such as pneumonia.

Remember I am talking about vitamin C complex and not ascorbic acid or ascorbates. Szent-Gyorgyi and others discovered that synthetic vitamin C (any form of ascorbic acid or ascorbate) will not cure scurvy. In his book, ***Oxidation***, he says that he received a letter from an Austrian colleague who was suffering from severe vascular hemorrhage and wanted to try ascorbic acid. "*Possessing at that time no sufficient quantities of crystalline ascorbic acid, I sent him a preparation of paprika that contained much ascorbic acid and the man was cured by it. Later with my friend, St. Rusznyak, we tried to produce the same therapeutic effect in similar conditions with pure ascorbic acid but we obtained* *no response. It was evident that the action of paprika was due to some other substance present in this plant..."*

In a 1988 article, a citrus extract of vitamin C was reported to be 35% more absorbed than a single dose of 500 mg of ascorbate and was found to be more bioavailable.[137]

Deficiency Symptoms

The first symptom of vitamin C deficiency is fatigue. This occurs long before other symptoms including bleeding gums, joint pain, weakness, tendency towards bruising, weight loss, and tendency to hemorrhage. Signs of vitamin C deficiency include depressed hemoglobin, defects in skeletal calcification, and problems resulting from the inability to metabolize the amino acids phenylalanine and tyrosine. As well, collagen, adrenal and spleen tissues develop pathological conditions with vitamin C deficiency.

Low levels of vitamin C are found in the following conditions: pregnancy, lactation, scurvy, during infectious diseases, congestive heart failure, kidney disease, liver disease, gastrointestinal disturbances, pupura, endocrine disorders, malignancies, burns, fractures, extensive surgery, and so on. Also, most children I have tested show vitamin C deficiency on the standard American (dead food) diet even though no clinical symptoms may be present.

In doing extensive determination of vitamin C via the Loomis 24-hour urinalysis , I have found that the two most common groups of people (other than children) who have extreme vitamin C deficiencies are smokers and people who take daily megadoses of ascorbic acid or ascorbates. I consider the Loomis urinalysis determination of vitamin C (via titration of a blue dye with urine to a colorless end point) an excellent and rapid method to determine vitamin C deficiency. This is based on the fact that vitamin C is a threshold substance - it is not excreted by the kidney until the blood level of ascorbic acid exceeds a certain value, which in turn depends upon the degree of saturation of the body tissues.

The fact that megadosers of ascorbate show vitamin C anemia is compelling clinical evidence that taking ascorbic acid alone without the other members of the vitamin C complex on a daily basis will lead in time to a vitamin C deficiency. This condition (vitamin C anemia) can be reversed rapidly, usually within one to four weeks, by giving the vitamin C complex, from raw foods or supplements derived from foods high in vitamin C.

Some Functions of Vitamin C Complex
(not ascorbic acid)

Essential to health and integrity of endothelial and connective tissues(collagen), adrenal and spleen tissues
Improves oxygen carrying capacity of blood
Functions with vitamin D in calcium metabolism
Promotes leukocytic and phagocytic activity, essential in fighting infection
Essential to proper development of teeth
Participates in regeneration of blood cells
Maintains proper blood clotting time
Combats effects of thyroid tachycardia

Source of food form vitamin C: Ness (*) has an enzyme combination (**#11**) containing plant enzymes plus food sources of bioflavanoids, acerola and rose hips. The enzymes digest the food so a much lower dose of the vitamin C food is required. Standard Process (*) makes a vitamin C complex called Cataplex C containing veal bone extract, adrenal glandular and buckwheat leaf juice. You can get a wonderful whole food form of vitamin C in a health food store from Bioforce. It is called Bio-C and each chewable tablet contains 100 mg of vitamin C from acerola and sea buckthorn extract in a fruit extract base. No ascorbic acid has been added to any of these products.

Food Sources of vitamin C complex: Vitamin C is the most difficult of the vitamins to get and is oxidized so easily that it disappears in storage of citrus fruits, vegetables, and upon being exposed to artificial lights such as the type at the supermarket. In decreasing order, the major food sources of vitamin C include: acerola, peppers (red chili, red and green sweet), guavas, kale leaves, parsley, collard leaves, turnip greens, and other leafy green vegetables, broccoli, and brussels sprouts, citrus fruits, berries, melons, tomatoes, potatoes and raw cabbage. Potato is a good source because in the potato the vitamin C is combined with a protein molecule. Baking the potato destroys less vitamin C than boiling it. Milk is also a good source of vitamin C unless it is pasteurized. Irradiation, the latest promoted method of food preservation, also destroys vitamin C.

Use of Ascorbic Acid as a Chemotherapeutic Agent During Acute Infection

Linus Pauling wrote a book on ***Cancer and Vitamin C,*** in which he talks about the use of ascorbic acid and sodium ascorbate, not vitamin C complex, which is one reason why people equate ascorbic acid with the vitamin C complex. Much research has been reported on the use of high doses of ascorbic acid/ascorbate, both intravenous and orally, in the treatment of acute infectious diseases and malignant degenerative states. Ascorbic acid becomes an effective pharmacologic agent when used in megadoses under these conditions. We should remember that the metabolism changes drastically during such periods. For example, during a cold, not only vitamin C needs to be increased, but the body utilizes more zinc than normally required which leads to loss of taste during the bout with the cold if extra zinc is not supplied. Cancer patients develop an ability to metabolize amounts of selenium that in a healthy person would be lethal. These transient metabolic shifts justify megadoses of nutrients not normally needed in these amounts on a daily basis.

In response to a Mayo Clinic study [138] which denounced ascorbic acid as worthless in cancer treatment, Pauling and Cameron stated that the Mayo Clinic study was flawed on three counts. **No. 1**: Urine tests for some of the patients receiving placebos showed abnormally high levels of ascorbate indicating that they were taking it without the researcher's knowledge. **No. 2**: Ascorbic acid was given for an average of ten weeks in the Mayo Clinic study while Pauling used the vitamin indefinitely. **No. 3**: Pauling says that the Mayo Clinic ignored the rebound effect which occurs when ascorbic acid is suddenly stopped. The resulting depression of the immune system may stimulate tumor growth. This illustrates that scientists may design experiments which reflect their bias rather than the truth.[139,140]

A potent antioxidant, ascorbic acid is is also a natural chelating agent (helps detoxify heavy metals).

Recently, many forms of ascorbic acid salts (ascorbates) have been introduced. Buffered ascorbic acid (ascorbate) is less irritating to the mucosal lining in some people than the acidic form. The latest ascorbate on the market is called Ester-C which contains ascorbic acid bonded to various minerals (calcium, magnesium, potassium, sodium and zinc). Some forms of Ester C contain bioflavanoids as well. Ester- C has been shown to be readily absorbed and retained longer by body tissues than other forms of ascorbate or ascorbic acid. So when one needs high doses of ascorbic acid, Ester-C would probably be a good choice. Remember, however, no form of ascorbic acid should be confused with the vitamin C complex described earlier, nor should their physiologies be considered identical.

Caution: Overdose symptoms: irritation of the gastrointestinal mucosa, flatulence, diarrhea, tendency towards kidney stones in some people, and interference with normal mineral metabolism. I do not recommend any form of ascorbic acid in large doses except during acute stages of infection.

Results of Vitamin C Deficiency

Fatigue, physical weakness
Joint pain
Decreased resistance to infection
Bruising easily
Weight loss
Shortness of breath
Rapid respiration
Rapid heart beat
Tendency to hemorrhage
Reduced hemoglobin
Hypertrophy and reduced adrenal secretion
Hypertrophy of thyroid (toxic goiter)
Decrease weight of spleen, liver, stomach and intestines
Increased tooth decay, gum disease
Increased vascular and cardiovascular disease
Tendency to peptic/duodenal ulcers
Friability of bones

CHAPTER 23: EXTERNAL PROTECTION

Low Level Radiation in your Environment: How to Find it and What to do about it

Because of the demand for more information on the effects of low level radiation, new companies are specializing in measuring radiation and chemical pollution in your home or workplace. I contacted one such company, called Safe Environments and was visited by Mr. John Banta (A), who came to my home with several thousand dollars of equipment to measure ionizing and non-ionizing radiation, geopathic stress zones, and radon levels. He found a radioactive rock in my living room, and a possible geopathic stress zone in my children's bedroom. Also he measured electromagnetic fields around my dimmer switches and photoelectric switches (that turn a night light on when it's dark and off when it's light). Although my TV and computers were safe at the operator's distance, John discovered that the wiring in my house was out-of-phase. Only one plug in the entire house was in-phase and grounded.

An out-of-phase circuit results when the hot (black) and neutral (white) wires are reversed, which is not an uncommon occurrence. This not only decreases the efficiency and lifetime of the appliances plugged into it, but also creates a much larger electromagnetic field than correctly wired circuits. He could measure this electromagnetic field even when the appliance was turned off but still plugged into the out-of-phase circuit due to an electric potential from the hot (black) wire. I had recently purchased a new copier which had never operated well and my problems were solved when we plugged it into a correctly wired and grounded outlet.

The worst place for electromagnetic fields is the bedroom because there are sometimes many appliances there, including a radio, TV, dial face clock, and an electric blanket, etc. Electric blankets are especially bad because they produce relatively strong fields and people lie under them for long periods of time. The same can be said for electric mattress pads and waterbed heaters. This is especially dangerous for the unborn child. Studies by Ruth Wertheimer and others have shown an abnormal increase in spontaneous abortion among pregnant women sleeping under electric blankets during the colder months of the year. Paul

Brodeur has a whole chapter on electric blankets in his new book, ***Currents of Death.*** The best thing to do is to unplug appliances near your bed and plug whatever you need into outlets six feet away from your bed.

Aqua Queen, available from Baubiologie Hardware (A) has introduced the first magnetically balanced waterbed heater which decreases the magnetic field from the conventional dose of 30-60 milligauss to less than one milligauss in the area of the sleeper.

Safe Environments plans to open offices in major cities across the United States. In the Los Angeles area, there is a similar company called Environmental Testing and Technology. There will probably be many more of these companies in the near future.

In February 1990, John and Trisha Banta released an excellent video entitled ***Current Switch,*** through their company, Baubiologie Hardware (A), which shows how to detect, reduce or eliminate electromagnetic pollution in the home. This 68-minute color video gives step-by-step information for eliminating or reducing electromagnetic fields from electric blankets, computers, electronic equipment (stereos, TV's, VCR's, clocks, radios), microwave ovens, high tension lines, microwave transmitters, radar and household wiring. It gives simple instructions for making your own inexpensive electromagnetic screening device, shows how to locate household wiring within the walls, explains how to replace dimmer switches, demonstrates identifying wires that are out of phase, and shows how to wire new or remodeled homes for maximum electromagnetic safety. It helps you to determine what corrections you can make and when to call in a master electrician, and helps you choose home locations for maximum electromagnetic safety.

Many new products for increasing environmental safety are emerging as the risks of electromagnetic pollution become common knowledge. For example, Baubiologie Hardware and other distributors are marketing an inexpensive ($100) meter available for measuring non-ionizing radiation including radiowaves, micro-waves, radar and low level electric and magnetic fields. Meters which measure ionizing radiation are much more expensive than this. You can get a catalog of wonderful "safe environment" products from Baubiologie Hardware and other similar sources.

To give you one more example, there is an inexpensive wallet sized UV detector card available which tells how much ultraviolet radiation is present. It can be used to test the efficacy of your sunglasses or to determine how strong your sun-screen cream should be. This card is marketed by Optiwear in Syracuse, New York. Talk to its inventor, Elliot Giles (A), for more information.

Natural Substances which Provide External Protection

The substances below are recommended for their ability to absorb and/or neutralize radiation. Very little research is available on some of these items. I have listed all the references I could find.

1. Crystals: Marcel Vogel (A) has spent many years studying the healing and protective effects of crystals on living organisms and on liquids, specifically water and wine. In a recent interview with Dr. Vogel, he told me that certain crystals can absorb radiation. This can be measured. For example smokey quartz turns this color from absorbed radiation. He is developing crystals which will act as radioprotective shields.

Dr. Vogel has completed a book on his research on crystals which will be published this year by Ballantine Publishers. It is called ***Marcel Vogel: on Crystals.***

2. Montmorillionite clay: The Engineering Research and Development Laboratories (Fort Belvoir, Virginia) wrote a paper entitled, "Decontamination of Radioactively Contaminated Water by Slurrying with Clay." The report presents data on the decontamination of radioactive water with montmorillionite clay.

3. Green plants and other chlorophyll sources such as barley or wheat grass: will absorb radiation similar to montmorillionite clay. A few blades of wheat (barley) grass will absorb heavy metals, fluoride and radiation from water. Similarly, green plants absorb pollution, odors, and radiation so they are more than just pretty decorations in your home. A plant on top of a TV, for example, not only looks lovely but absorbs radiation from the TV. This is described in all of Betsy Russell Manning's books on wheat grass, silver-mercury fillings and candidiasis.

Notes on Green Plants

Green plants do far more than inspire carbon dioxide and expire oxygen for us to breathe. They work in air, water and inside our bodies to cleanse, purify and heal. I have described the profound healing effects of green foods on our bodies in Chapter 17. Here I want to give you a few more thoughts on how green plants detoxify air and water. First, let's talk about how plants cleanse the air.

Energy efficient houses have not only increased radon but other common household chemicals. This includes new products such as pressed wood furniture and fiberboard (which emit trace levels of organic chemicals) plus the many common household chemicals such as cleaners, insecticides, air freshners and hair sprays.

In 1983, an EPA study found over 350 volatile organic compounds in one Washington D.C. nursing home. The worst pollutants are radon, tobacco smoke, pathogenic (disease-causing) micro-organisms, volatile organic compounds, by-products of coal, oil and wood burning, asbestos, and combustion gases such as carbon monoxide, nitrogen dioxide and particles.[141]

The effects of this pollution include irritation of eyes, nose, throat, lungs, damage to the nervous system, kidneys liver and heart, impaired development in children and increased risk of cancer. [141]

NASA also studied indoor air pollution after identifying 107 volatile organics inside the Skylab spacecraft.[141]

Should we all move outside into our tents? Again, the answer is nature. Plants absorb radiation, heavy metals, fluoride, microbes, etc. from both air and water. Below is a list of common plants known to absorb air pollutants. Three air pollutants-formaldehyde, benzene and trichloroethylene - were studied in particular because they are the most common organics found indoors and have been implicated in many cases of "sick-building syndrome" (I wonder why they call it "sick-building" and not "sick-people-who-work- in-building" syndrome ?!?).

We do not know how plants remove pollutants but that they do, with or without their foliage. How many plants are needed? No one know for sure although estimates have been made. My advice is to minimize sources of pollution and "go for it" with your plants (i.e. the more plants the more beautiful!).[141]

Now, what about how plants cleanse the water? This is a whole topic of research. I will briefly mention one research paper to illustrate that plants are not only beautiful but necessary for planetary life. In particular, this research paper emphasized one special plant called Schoenoplectus lacustris which was found to detoxify toxic heavy metals, phosphorus (a polluting by-product of detergents), and extremely toxic organic compounds such as phenols and cyanide.[142]

In bacteriological and biological respects, the ability of plants to purify sewage has totally surpassed any sewage-treatment plant performance. For example, in 48 hours, ten species of plants eliminated E. coli, a common pathogen by 99%. In this experiment, common plants such as reeds and lilies were tested as well as Shoenoplectus.

Let me end with a quote from Frederic Vester that I found in the above article, who compares our modern and inefficient sewage treatment centers with plant (biological) purification. He says that "*this system has existed now for some four thousand million years and it is certainly advisable to learn from a business that has not yet gone broke after such a long time.*" He queries, "*can plants teach our captains of industry? These plants turn over 200 thousand million tons of organic material and produce 100 thousand million tons of oxygen and process metals at a rate of many thousands of million tons year after year. This is an imposing turnover of energy and matter with as high as 98% efficiency.*" [143]

I leave it to you to imagine how much better it would be if we cooperated with nature in our efforts to create a better environment for ourselves and our children.

Air Cleaning Houseplants

Aloe
Reed palm
Spider plant
Pot mum
Janet Craig dracaena
Dragon tree
Warneckeii dracaena
Golden pothos
African daisy
English ivy
Banana tree
Lacy tree philodendron
Heart-leaf philodendron
Peace lily
Snake plant

CHAPTER 24:
ANSWERS TO FREQUENT QUESTIONS

Should you run out and take all this stuff? Which ones, and for how long. How much should you take? What do I (and my family) take? How do you tell what you need? Are synthetic vitamins okay? Which ones? Below are my thoughts on these questions.

I used to believe that whatever nutrient deficiency resulting from eating junk foods, processed foods, etc. could be corrected by simply taking that nutrient. I no longer believe this. In fact, I think it's a total fantasy to think that you can eat junk foods and supply the missing nutrients by taking synthetic vitamins and minerals.

Secondly, to repeat what I've said in Chapter 21, I do not believe that taking large doses of any single nutrient is healthy because it causes an imbalance in the other nutrients with which it works. It's not that I am against vitamins and minerals. In general, except for very short term therapy, and during acute illness, I do not believe that it is healthy or wise to take high dose vitamins and minerals.

What about synthetic vitamins? There is a general rule that I use here. The more complicated the vitamin, the harder it is to synthesize. For example, natural vitamin E is very different from synthetic vitamin E though both have vitamin E activity. I prefer the natural form, or better yet, vitamins from foods. I have trouble believing that synthetic vitamins and minerals will substitute for natural foods.

My family program includes: Ness food enzymes; Biotec antioxidant enzymes; the green foods, specifically, barley grass and Blue Green Algae; and Vitalplex probiotics by Klaire Labs. Occasionally, I will use a low dose vitamin and chelated mineral supplement, especially if our family has been exposed to less than healthy foods, but my major emphasis is on enzyme nutrition.

Finally, remember your attitude, thoughts and feelings. They have a profound effect on your immune system. Despair, fear and depression are just as unhealthy as junk food.

1. Joseph Rotblat, *Science News,* 128 (15) 1986.
2. *East West Magazine,* June 1986.
3. Lowell Ponte, "The Menace of Electric Smog," *Reader's Digest,* March 1986, p. 123.
4. Paul Brodeur, "Radiation in Daily Life,"*US,* Dec. 1, 1986.
5. Abram Petkau, Ph.D., "Effect of 22Na+ on a Phospholipid Membrane," *Health Physics,* 22:239 1972.
6. Ernest J. Sternglass, Ph.D., "The Implications of Chernobyl for Human Health," , *Int. J. of Biosocial Research,* 8 (1): 7-36, 1986.
7. David L. Watts, M.D., "Prevalence of Lead in the Environment Threatens Children," *Health Freedom News,* Oct. 1985.
8. Associated Press, *San Leandro Daily Review,* May 1, 1986.
9. W.H. Beirwaltes et al., "Radioactive Iodine Concentration in the Fetal Human Thyroid Gland from Fallout," *J. Am. Med. Association,* 173: 1895, 1960.
10. Jim Sibbison, "Radiation and the Unborn," *Bestways,* Jan. 1984, p. 14 and references therein.
11. A. Stewart, J. Webb and D.A. Hewitt, "Survey of Childhood Malignancies," *Brit. Med. J.,* 1:1495-1508, 1958.
12. A. Stewart and G.W. Kneale, "Radiation Dose Effects in Relation to Obstetric X-Rays and Childhood Cancers," *Lancet,* 1:1185-1188, 1970.
13. Elsie M. Widowson et al., "Absorption, Excretion and Retention of Strontium by Breastfed and Bottle-fed Babies," *Lancet,* Oct. 29, 1960.
14. Robert Mendelsohn, M.D., *Health Express,*July 1981, p. 48.
15. John Gofman, "Ionizing Radiation: Some Concepts for the Dental Assistant," 3-part article in *The Dental Assistant,* Aug., Sept., and Oct. 1982. Adapted from Gofman's book, ***Radiation and Human Health,***1981.
16. Frank Schiavo, *San Jose Mercury,* May 9, 1986.
17. Associated Press, *San Francisco Chronicle,* May 4, 1986.
18. *Campaign California Newsletter,* 4th Qrtr., Vol. 2, No.4, 1987.
19. *Abalone Alliance, Radiation Bulletin,* vol. 4, no. 1, 1989.
20. *San Francisco Examiner,* Fri., June 2, 1989.
21. Harvey Wasserman, "Three Mile Island Did It," *Harrowsmith,* March 28, 1979.
22. Brian Jacobs, "The Politics of Radiation When Public Health and the Nuclear Industry Collide," *Greenpeace,* July/Aug. 1988.
23. *San Jose Mercury News,* Sept. 23, 1986.
24. S. Applegate, *San Jose Mercury News,* May 4, 1986.
25. Publication from the Committee for Nuclear Responsibility. (A)

26. *San Jose Mercury News*, May 22, 1986.
27. Lowell Ponte, "The Menace of Electric Smog," *Reader's Digest*, Jan. 1980.
28. Mike Langberg, "Nuclear Industry Tries for Comeback in 1990's'," *San Jose Mercury News*, Oct. 2., 1989.
29. Excerpts from *Acres, U.S.A.*, May 1984, reprinted in *Health Freedom News,* Oct. 1984.
30. Dr. Jazsel Barna, "Compilation of Bioassay Data on the Wholesomeness of Irradiated Food Items," Central Research Institute, H-1022, Budapest, Herman Otto ut15, Hungary, reprinted by Coalition for Alternatives in Nutrition & Healthcare (A).
31. J. Tevere MacFadyen, "Immortal Shrimp and Other Fruits of the Gamma Ray," *Harrowsmith*, Jan./Feb. 1986, p. 43.
32. "The Safety of Food Irradiation," *The Monthly Planet*, 320-G Cedar St., Santa Cruz, CA 95060, Oct. 1985.
33. Takahashi Kosei, "Irradiation of the Food We Eat, A New Danger," *AMPO*, 15 (2): 50, 1983.
34. *The Food Irradiation Alert!* Newsletter of the National Coalition to Stop Food Irradiation, Nov./Dec. 1985 - Jan./Feb 1986.
35. Bruce Swain, "The Three-Mile Island of Irradiation," *East West Magazine,* Oct. 1989.
36. Tom Johnson, "Ban Approved on Irradiated Food," *Newark Star-Ledger,* Feb. 3, 1987.
37. Paul Brodeur, "Annals of Radiation, Part I - Power lines; Part II - Electromagnetic Fields; Part III - Video Display Terminals," *New Yorker Magazine*, June 12, 19 and 26, 1989.
38. Andrew J. White, *Spotlight*, Oct. 19, 1981.
39. "Microwave Ovens: How Safe? No One Really Knows," *NHF Bulletin*, Oct. 1980.
40a. Thomas Hartmann, "Computers and Your Health," *East West Magazine*, May 1984. Discussion of FDA Study #81-8153, 1978, now listed as out of print.
40b. Kirk Johnson, "Test Your Computer for X-Rays," *East West Magazine*, October 1986.
41. *The Toronto Star*, Sept. 29, 1983.
42. C.R. Hirning and J.H. Aitken, Canadian Government's Radiation Protection Service, Ministry of Labour, *Health Physics*, 43 (5): 727, 1982.
43. Jeff Sorenson and Jon Swan, "VDT's The Overlooked Story Right in the Newsroom," *Columbia Journalism Review,* No date given but printed after 1984.

44. "Don't Sit too Close to the TV: VDT's CRT's and Radiation," Cold Type Organizing Committee, P.O. Box 40, Jerome Ave. Station, Bronx, N.Y. 10468.
45. Thomas Hartmann, "Computers and Your Health," *East West Journal,* May 1984.
46. Peter H. Lewis, *San Jose Mercury,* Oct. 13, 1985; *New York Times,* Sept. 3, 1985.
47. "VDT Controversy: Health Hazard or Annoyance?" *Health Freedom News,* Nov. 1983.
48. "VDT's Byte the Bullet," *East West Journal,* July 1984.
49. John Ott, *Townsend Newsletter for Doctors,* Issue #37, p. 127, May 1986.
50. John Ott, *Health Consciousness,* Vol, VIII, No. 1, Feb. 1987. POB 550, Oviedo, FL 32765. $24/6 issues.
51. John Ott, "Color and Light, their Effects on Plants, Animals and People," Part 2, *International Journal of Biosocial Research,* Special Subject Issue, Vol 8, 1986.
52. Dolores Mason, 526 Sanders, Indianapolis, IN 46203, (317) 636-2615.
53. Louis Slesin, ed., *VDT News,* POB 1799, Grand Central Station, NY 10163. $87 for bimonthly subscription.
54. Bill Paul, "IBM to Reduce Radiation from Future VDT Models," *Wall Street Journal,* Wed., Nov. 22, 1989.
55. Bill Paul, "Latest Study on VDT's Adds to Safety Fears," *Wall Street Journal,* Oct. 20, 1988.
56. Diane Leighland, "The SImple Relationship Between Trees, Cars, Oxygen and Ozone," Courtesy of the *Kama'aina News,* reprinted by the We Foundation, POB 1408, Makawao, HI 96768.
57. John Ott, POB 40004, Crescent Beach, Sarasota, FL 34242.
58. "Let There be Light," *Cancer Control Journal,* Double Issue, Vol. 3 (No. 6) and Vol . 4 (No. 1), Feb. 1976. See (A) for address.
59. Lowell Ponte, "How Artificial Light Affects your Health," *Reader's Digest,* Feb. 1981.
60. Ronald E. Kotzsch, "Bringing the Sun Indoors," *East West Magazine,* March 1989, p. 48.
61. Valerie Beral et al, "Malignant Melanoma and Exposure to Fluorescent Lighting at Work," *The Lancet,* Aug. 7, 1982, p. 290.
62. Robert Stern, "Malignant Melanoma and Exposure to Fluorescent Lighting at Work," *The Lancet,* Nov. 27, 1982, p. 1227.
63. *Food and Chem. Toxicology,* Vol. 22, p. 431-4376, 1984.
64. *San Jose Mercury News,* May 9, 1990.

65. Nathaniel Mead, "The Riddle of Radon," *East West Magazine,* July 1990, p. 64.
66. John Yiamouyiannis, Guest Editorial, "NIDR Study Shows No Relationship between Fluoridation and Tooth Decay Rate," *American Laboratory,* May 1989.
67. Bette Hileman, "New Studies Cast Doubt on Fluoridation Benefits," *C & E News,* May 8, 1989.
68. Harriett Fels, "Fluoridation Health Measure or Hoax," *East West Magazine,* Oct. 1989, p. 62.
69. James F. Scheer, "New Washington Scandal: the Fluoridation Issue," *Health Freedom News,* Mar./April 1990.
70. "Fluoride Found to Cause Cancer in Rats; Human Risk Discounted," *San Jose Mercury News,* Jan. 29, 1990.
71. L J. Hahn et al, "Dental Silver Tooth Fillings: a Source of Mercury Exposure," *The FASEB J.,* Vol. 3, Dec. 1989, p. 2641.
72. *Health Consciousness Magazine,* Roy Kupsinel, M.D., editor and publisher, POB 550, Oviedo, FL 32765. (407) 365-6681, $30/6 issues.
73. D. Eggleston, D.D.S., *J. Prostatic Dentistry,* May 1984.
74. *San Francisco Chronicle,* May 8, 1990.
75. Wright/ Gaby Nutritional Seminars, "Clinical Applications of Nutritional Biochemistry,"Jonathan Wright, M.D., Alan Gaby, M.D., 1983. Address: 6931 Fieldcrest Rd., Baltimore, M.D. 21215. (301) 764-3471.
76. G. Hems, "Sugar and Breast Cancer, " *Brit. J. of Cancer,* June 1978.
77. K.P. Porikos, T. B. van Itallie, "Diet Induced Changes in Serum Transaminase and Triglycerides Levels in Healthy Men. Role of Sucrose and Excess Calories," *Am. J. Med.,* 75: 624, 1983.
78. Woodrow C. Monte, Ph.D., "Aspartame: Methanol and the Public Health," *J. of Applied Nutrition,* Spring 1984, pp. 42-54.
79. Barbara Mullarkey, "Will the New Sweetner Aspartame Leave America Bitter?"*Vegetarian Times,*Sept. 1984, p. 47.
80. "Research on NutraSweet Reveals Irregularities in Mice," *Times Tribune,*Sept. 7 1985.
81. *Clinical Nutrition Newsletter,* vol. 2 (12) Dec. 1982.
82. *Health Care Letter,* vol. 1 (2), Spring 1990, POB 87, Bryn Mawr, PA 19010.
83. Alan Immerman, "Evidence for Intestinal Toxemia - an Inescapable Clinical Phenomenon," *The ACA Journal of Chiropractic,* Vol 13: S25-S36, April 1979.

84. M. Weiss, *West Magazine, San Jose Mercury*, May 27, 1990.
85. Peter Rothschild, M.D., "Enzyme Therapy in Immune Complex and Free Radical Contingent Diseases," State Academy of Medicine, Tamaulipas, Mexico, Published by University Labs Press, A Division of University Labs, Honolulu, HI 96826.
86. Stephen Levine, Ph.D. and J. Frackelton, M.S., *J. Orthomolecular Psychiatry*, 3rd Quarter, vol 12 (3): 166, 1983.
87. H. Sies and A. Wendel, ***Functions of Glutathione,*** Springer-Verlag, New York, 1978.
88. Henry J. Younes, D.O., Cardiovascular Anesthesiology, "Preoperative Nutritional Support in Cardiac Surgical Patients," Huntsville Hospital, Huntsville, AL 35801. Office: 925-D Franklin Street, Huntsville, AL 35801.
89. *Berkeley Wellness Letter,* University of California, Berkeley, CA, April 1988.
90. K. Kubota, Y. Matsuoka & H. Seki, "Isolation of Potent Anti-Inflammatory Protein from Barley Leaves," *Japanese Journal of Inflammation,* Vol. 3, No. 4, 1983.
91. K. Kubota and Y. Matsuoka, "Effect of Chronic Administration of Green Barley Juice on Growth Rate of Mice." Presented at Congress of Pharm. Soc. of Japan, Sendai, 1984. Faculty of Pharmaceutical Sciences, Science Univeristy of Tokyo, 12 Ichigayafunagawara-machi, Shinjuku-ku, Tokyo.
92. *Atherosclerosis,* 25: 199-204, 1976.
93. John Prudden and Leslie Balassa, Seminars on Arthritis and Rheumatism, vol III, no. 4: 287, 1974.
94. Betty Kamen, Ph.D., "Algae, Packaged Sunlight," *Let's Live,* Nov. 1985.
95. Hideo Nakayama, "Chlorella, the Algae that Cures," *Health World,* March/April 1989.
96. K.M. Shahani , Ph.D. and A.D. Ayebo, Ph.D., *The American Journal of Clinical Nutrition,* 33: 2448-57, Nov. 1980.
97. William C. Douglas, ***The Milk of Human Kindness is Not Pasteurized,*** p. 254. See (B) for address.
98. Ibid., p. 253.
99. Alan R. Gaby, M.D.. *Townsend Letter for Doctors,* Feb./March 1990, p. 82.
100. Edouard Brochu, *Townsend Letter for Doctors,* Feb./March 1989, p.99.
101. Leon Chaitow, "Food Hygiene.", *Health Consciousness Magazine,* Oct. 1988.
102. Daniel P. Puzo, *San Jose Mercury,* May 4, 1990.

103. *East West Journal,* July 1984.
104. Johannes Kuhl, ***Check Mate for Cancer,*** Viadrina Vertag, A., Trowitzch, Braunlage, Harzburger Str. 6, W. Germany.
105. Tom Valentine, *Spotlight*, June 2, 1986.
106. National Research Council Report: ***Diet, Nutrition and Cancer,*** National Academy Press, 2101 Constitution Ave., N.W., Washington, D.C., 1982.
107. "Stop Rectal Cancer," *Cancer Control Journal*, Vol. 6, No. 7-12, July-Dec. 1978.
108. R. Mendelsohn, M.D., *Whole Life Times,* Jan.-Feb. 1984.
109. Louise Tenney, *Cancer News Journal,* Spring 1982.
110. Alan Bock, *Health Freedom News,* Vol. 5, No. 4, April 1986.
111. K. Donsback, Ph.D. & R. O. Brennan, D.O., ***Dr. Donsbach Tells You What You Always Wanted to Know About Superoxide Dismutase,*** International Institute of Natural Health Sciences, 7422 Mountjoy Dr., Huntington Beach, CA 92648.
112. ***1984-1985 Yearbook of Nutritional Medicine, First Edition,*** Jeffrey Bland, Ph.D., Editor, Keats Publishing Co., Inc., New Canaan, CT 06840.
113. F. W. Sunderman, *Fed. Proc., 37: 40-46, 1978.*
114. Stephen A. Levine, Ph.D. & Jay Parker, "Selenium and Human Chemical Hypersensitivities, Preliminary Findings," *Journal of Biosocial Research,* 3 (1): 44-47, 1982. Address: POB 1174, Tacoma, WA 98401.
115. G. Fahy, Ph.D., *Anti-Aging News,* Vol. 3, No. 7, July 1983.
116. *Biomedical and Clinical Aspects of Coenzyme Q,* K. Folkers & Y. Yamamura, Editors, Vol. 1-4, Elsevier Science Publishers, Amsterdam, 1984.
117. G.P. Littarru at al., "Deficiency of CoQ10 in Gingival Tissue from Patients with Periodontal Disease," *Proc. National Academy of Science,* 68: 2332, 1971.
118. Robert M. Downs, D.C.& Alice Van Baak, "Coenzyme Q10," *Bestways,* June 1987.
119. D. F. Horrobin, M.D., *J. Holistic Medicine,* 3 (2): 118, 1981.
120. E. Signel, "Cancerostatic Effects of a Vegetarian Diet," *Nutrition and Cancer,* 4: 285, 1983.
121. R. M. Hicks, *Proc. Nutr. Soc.,* 42: 83-93, 1983.
122. Jonathan V. Wright, M.D., *Am. Holistic Medical Association Newsletter,* Oct. 1982.
123. Y-mour Anon, *Nutr. Rev.,* 40 (9): 265-8, 1982.

124. Durk Pearson & Sandy Shaw, ***Life Extension, A Practical Scientific Approach,***Warner Books, 75 Rockefeller Plaza, New York, NY 10019, 1982.
125. A. Lopez, M.D. & B. LeGardeur, *Clinical Research,* Vol. 30, No. 2, 1982.
126. The National Research Council, *Geochemistry and the Environment,"* Vol. III, Washington, D.C: National. Academy of Sciences, 1978.
127. Morton Walker, D.P.M., *Townsend Newsletter for Doctors,* #52 8711, p. 312, Nov. 1987. Correspondence: 484 High Ridge Rd., Stamford, CT 06905.
128. K. Donsbach, Ph.D., ***Glandular Extracts,*** Address in footnote 111.
129. Ivan Popov, M.D., *Journal of the Int. Acad. of Preventive Medicine,* July 1977.
130. Simon Martin, Editor, *Here's Health,* Victory House, Leicester Place, London WC2H 7BP, Tel: 011-4471-437-9001.
131. A.G. Schwartz et al., "DHEA, an Anti-Obesity and Anti-carcinogenic Agent," *Nutrition and Cancer,* 3(1): 46-53, 1981.
132. *Nutrition News,* 1989, POB 55279, Riverside, CA 92517.
133. Pycnogenol: U.S. Patent No. 4,698,369, granted to Dr. Jack Masquelier of the University of Bordeaux.
134. Richard P. Murray, D.C.& W. J. McCormick, M.D., *The Clinical Nutrition Newsletter,* National Academy of Research Biochemists, 1983. Address: 1323 Eagle Point Circle, Biloxi, M.S. 39532.
135. Richard Murray, D.C., *Biomedical Nitty-Gritty,* Biomedical Health Foundation, Inc., 69 West Granada Blvd., Ormond Beach, FL 32074.
136. Howard Loomis, D.C., 21st Century Nutrition, "Seminar Series on Enzyme Nutrition." See (A) for address.
137. *Am. J. Clinical Nutr.,* 48: 601, 1988.
138. *New England Journal of Medicine,* 312: 137, Jan 17, 1985.
139. *Health Freedom News,* Vol. 4, No. 4, April 1985.
140. *East West Journal,* April 1986.
141. Janet Marinelli, *San Jose Mercury News,* May 17, 1990.
142. Kathe Seidel, *Garten and Landschaft,*1/78 (In English, too)

Companies, Organizations and People Providing Environmentally Safe Products and Food Supplements and Information on these Topics.

Anti-nuclear groups:

*The Abalone Alliance:** 2940 16th St, #310, San Francisco, CA 94103; (415) 861-0592. Publishes *Radiation Bulletin.*
***Committee for Nuclear Responsibility, Inc.**, POB Box 11207 San Francisco, CA 94101.
***Food and Water, Inc.**,Michael Colby, 225 Lafayette Street, Suite 612, New York, NY 10012; (212) 941-9340.
***National Coalition to Stop Food Irradiation**, P.O. Box 59-0488, San Francisco, CA 94159. (415) 626-2734.
***National Coalition to Stop Food Irradiation**, Joel Hill, POB 2384, Stanford, Ca 94309

Antioxidant enzymes:

***Biomed Foods**: Zane Baranowski, 597 Glenwood Cutoff, Scotts Valley, CA 95066; (800) 456-4959.
***Biotec Foods of Hawaii**: POB 3266, 1215 Center St., Honolulu, HI 96816. (800) 331-5888 inside and outside California.
***Food Doctor**: Scott Springer, 208 East Alma, Mount Shasta, CA 96067; (916) 926-4301 (in CA); (800) 327-1956 (U.S.).

Environmental Protection/Health Oriented Groups, Companies and Products:

***Baubiologie Hardware:** See "Current Switch".
***Cancer Control Society:** L. Rosenthal, Co-founder, 2043 N. Berendo Street, Los Angeles, CA 90027; (213) 663-7801.
***Coalition for Alternatives in Nutrition & Healthcare (CANAH)**, POB B-12. Richland Town, PA 18955.
***Current Switch:** by John and Trisha Banta, a 68-minute video on how to reduce/eliminate electromagnetic pollution in your home, $49.95 plus tax and shipping from Baubiologie Hardware L.P., 207B 16th Street, Pacific Grove, CA 93950; (408) 372-8626.
***Greenpeace:** 1436 U Street NW, Washington, DC 20009. Publishes bimonthly *Greenpeace Magazine.*
***Labor Occupational Health Program (LOHP),** 2521

Channing Way, Berkeley, CA 94720; (415) 642-5507. Has library on occupational hazards including VDT information.

***National Health Federation:** Box 688, Monrovia, CA 91016. Publishes *Health Freedom News Journal.*

***Optiwear:** makes UV detector card. Contact Elliott Giles, 5215 Dwane Drive, Syracuse, New York 13066; (315) 637-8504 (office and fax); (315) 463-0055.

Food Enzymes:

***21st Century Nutrition:** POB 640, Forsyth, MO 65653; (800) 637-7774; FAX (417) 546-6433.

***NESS (Nutritional Enzyme Support System):** POB 249, Forsyth, MO 65653. Customer Service: (800) 6333-7478, Order desk: (800) 637-7893.

Foods with Special Purposes:

***Antioxidant enzymes:** listed as such.

***Blue Green Algae** (called **Klamath Blue Green Algae**): same address as Food Doctor above.

***Blue Green Algae (Cell Tech):** 1300 Main Street, Klamath Falls, OR 97601. Has video tape too.

***Eggs, vaccinated**: San Pasqual Academy Farms, Rt 1, POB 890, Escondido, CA 92025.

Food Enzymes: listed as such.

***Green Foods (Barley Grass):** 620 Maple Ave., Torrance, CA 90503; (800) 777-4430; (213) 618-0678.

***Klaire Labs** (Probiotics): P.O. Box 618, Carlsbad, CA 92008; (619) 744-9680.

***Rejuvenative Foods** (lactic acid vegetables): Evan Richards, POB 8464, Santa Cruz, CA 95061; (408) 462-6715.

***Sun Chlorella Co:** 4025 Spencer St. #103, Torrance, CA 90503; (800) 537-0088.

***Spicer's International** (high fiber foods for diet and health) : POB 99, Oglesby, IL 61348; (815) 883-3506.

Full Spectrum glasses:

***Wesley-Jessen Company** (full spectrum contact lenses): 400 West Superior, Chicago, IL 60601; (312) 751-6200.
***Younger Optics** (full-spectrum ultraviolet transmitting eyeglasses): 3788 Broadway Place, Los Angeles, CA 90007; (213) 232-2345.

Ozone depletion:

***The We Foundation:** POB 1408, Makawao Hawaii 96768.

People involved in pioneering research on radiation and other environmental issues:

***Banta, John** (environmental detection of and protection from low level radiation): See Baubiologie Hardware.
***Baranowski, Zane** (Biotec antioxidant enzymes) 597 Glenwood Cutoff, Scotts Valley, CA 95066, 1-800-456-4959.
***Fahey, Jim** (developed Biotec antioxidant enzymes): See Biotec Foods of Hawaii under antioxidant enzymes.
***Gofman, John, M.D., Ph.D.** (co-discoverer of Uranium 233, research on effects of low level ionizing radiation, founder of Committee for Nuclear Responsibility): Contact via the committee
***Livingston-Wheeler, Virginia, M.D.** (developed anti-cancer vaccine): Livingston-Wheeler Medical Clinic, 3232 Duke St., San Diego, CA 92110; (619) 224-3515 or 692-2456.
***Ott, John, Ph.D.** (effects of light on health, full-spectrum light): POB 40004, Crescent Beach, Sarasota, FL 34242.
***Rothschild, Peter J., M.D.**, (antioxidant enzyme chemistry, chemistry of radioprotection): POB 1290, Eureka MT 59917.
***Shahani, Khem, Ph.D.**(probiotics): University of Nebraska, Food, Science and Technology Dept., 115 Filley Hall, Lincoln, NE 68583; (402) 472-2815.
***Sternglass, Ernest J., Ph.D.** (radiation research): 11 Riverside Dr., Apt 16 VE, New York, NY 10023.
***Vogel, Marcel, Ph.D.** (use of crystals in healing): Lab: PRI, 1725 Little Orchard St., Unit C, San Jose, CA 95125, (408) 279-2291; Home: (408) 292-2920.
***Yiamouyiannis, John, Ph.D.** (effects of fluoridation on health): See The Safe Water Foundation below for address.

VDT Screens and low radiation items:

***Aqua Queen** (magnetically balanced waterbed heater): Available from Baubiologie Hardware.
***Ecosystem (Somascreen)** 1520 Eastlake Ave. East, #103, Seattle, WA 98102; Orders: (206) 328-3368
***NoRad Corporation,** 1549 11th Street, Santa Monica, CA 94041; (213) 395-0800; (800) 262-3260.
***Safe Computing (liquid crystal monitor)** 368 Hillside Ave., Needham, MA 02194; (800) 222-3303; (617) 444-7778.

Water:

***Sea salt/soda bath**: School of Scientific Nutrition, 1605 Coal Ave. S.E., Albuquerque, NM 87106.
***Water:** The Safe Water Foundation, 6439 Taggart Road, Delaware, Ohio 43015. Founder: John Yiamouyiannis.

Books

Airola, Paavo, Ph.D., N.D., ***Dr. Airola's Handbook of Natural Healing, How to Get Well,*** Health Plus Publishers , POB 22001, Phoenix, AZ 85028.

Becker, Robert, M.D., ***Cross Currents,*** Jeremy P. Tarcher, Inc., Los Angeles, 1990.

Becker, Robert, M.D. and Gary Selden, ***The Body Electric,*** William Morrow, New York.

Bland, Jeffrey, Ph.D., ***Bioflavanoids,*** Keats Publishing, Inc., Good Health Guides, Box 876, New Canaan, CT 06840.

Bland, Jeffrey, Editor, 1984-85 ***Yearbook of Nutritional Medicine, First Edition,*** Keats Publishing Co., Inc., New Canaan, CT 06840.

Bragg, Paul, N.D., Ph.D. & Patricia Bragg, N.D., Ph.D., ***The Shocking Truth About Water,*** Health Science, POB 7, Santa Barbara, CA 93102, 1981.

Brodeur, Paul, ***The Zapping of America: Microwaves, their Deadly Risk, the Coverup,*** W.W. Norton & Co., 500 5th Avenue, New York 10036.

Brodeur, Paul, ***Currents of Death,*** Simon and Schuster, Rockefeller Center, 1230 Ave. of the Americas, New York, NY 10020, 1989.

Clark, Linda, M.A., ***Handbook of Natural Remedies for Common Ailments,*** Devin-Adair Co., Old Greenwich, CT, 1976.

Deal, Sheldon C., D.C., ***New Life Through Nutrition,***1974.

Donsbach, K.W., Ph.D., R. Welch, M.D., ***Cancer, Cause and Control,*** International Institute of Natural Health Sciences, 1981., 7422 Mountjoy Drive, Huntington Beach, CA 92648. 1981.

Donsbach, K.W., Ph.D., D.C., N.D., ***Dr. Donsbach Tells You***

What You Always Wanted to Know About Drinking Water.

Douglas, William Campbell, M.D., ***The Milk of Human Kindness is Not Pasteurized,*** POB 1568, Clayton, GA 30525; (800) 227-6269

Fredericks, Carlton, Ph.D., ***Nutrition Guide for the Prevention and Cure of Common Ailments and Diseases,*** Simon and Schuster, Inc., 1230 Avenue of the Americas, New York, NY 10020, 1982.

Gofman, John W., M.D., Ph.D., ***Radiation and Human Health,*** 1981, from Committee for Nuclear Responsibility, Inc., P.O. Box 11207, San Francisco, CA 94101. Other books are available.

Hagiwara, Yoshihide, M.D., ***Green Barley Essence,*** Keats Publishing Co., New Canaan, CT, 1986.

Howell, Edward, M.D., ***Enzyme Nutrition, the Food Enzyme Concept,*** Avery Publishing Group, Inc., Wayne, NJ, 1985. Available from 21st Century Nutrition (A).

Howell, Edward, M.D., ***Food Enzymes for Health and Longevity,*** Omangod Press, POB 64, Woodstock Valley, CT 06282, 1980. Available from 21st Century Nutrition (A).

Huggins, Hal A., D.D.S., ***It's All In Your Head,*** HAH Publications, 1985, POB 2589, Colorado Springs, CO 80901; (303) 473-6324; (800) 331-2303.

Kamen, Betty, Ph.D. & Si Kamen, ***Osteoporosis,*** , Pinacle Books, NY, 1984. Nutrition Encounter, POB 689, Larkspur, CA 94939; (415) 924-6409.

Kamen, Betty, Ph.D., ***Germanium, a New Approach to Immunity,*** Nutrition Encounter, POB 689, Larkspur, CA 94939, 1987. Other books: ***Nutrition in Nursing,*** and more.

Kenyon, Julian N., M.D., N.D., ***Modern Techniques in Acupuncture,*** Thorsons Publishers, England, 1985. May be ordered from China Books, 24th St., San Francisco, CA.

Kuhl, Johannes, ***Check Mate for Cancer,*** Viadrina Vertag, A., Trowitzch, Braunlage, Harzburger Str. 6, W. Germany.

Kupsinel, Roy, M.D., ***A Patient's Guide to Mercury Amalgam Toxicity,,*** P.O. Box 550, Oviedo, FL 32765, 1984.

LaFavore, Michael, ***Radon, the Invisible Threat,*** Rodale Press, Emmaus, PA, 1987.

Levine, Steve, Ph.D. & P. Kidd, Ph.D., ***Antioxidant Adaptation,*** Biocurrents Press, 400 Preda St., San Leandro, CA 94577, 1986.

Licata, Vincent, ***Comfrey and Chlorophyll,*** Continental Health Research, 300 N. Broadway, Santa Ana, CA 92701, 1971.

Livingston-Wheeler, Virginia, M.D., ***Food Alive, a Diet for Cancer and Chronic Diseases,*** Livingston-Wheeler Medical Clinic, 3232 Duke St., San Diego, CA 92110, 1977, (619) 224-3515, 692-2456. One of a four book series.

Long, Ruth Yale, Ph.D., ***Crackdown on Cancer,*** Nutritional Ed. Assn., POB 20301, Houston, TX 77225; (713) 665-2946.

Mindell, Earl, Ph.D., ***Earl Mindell's Quick and Easy Guide to Better Health,*** Pivot Health Books, Keats Publishing, Inc., Box 876, New Canaan, CT

National Research Council Report, ***Diet, Nutrition and Cancer,*** National Academy Press, 2101 Constitution Ave., N.W., Washington, DC, 1982.

Neustaedter, Randall, C.A., ***The Immunization Decision, a Guide for Parents,*** Homeopathic Educational Service, 2124 Kittredge Street, Berkeley, CA 94704.

National Research Council Report: ***Geochemistry and the Environment,*** Vol. III. Washington, DC: National Academy of Sciences, 1978.

Ott, John, ***Color and Light,*** Part II, *Int. Journal of Biosocial Research, Special Subject Issue,* Vol. 8, 1986.

Ott, John N., ***Health and Light***, Devin-Adair Co., One Park Ave., Old Greenwich, CT 06870, 1973.

Page, Melvin E., D.D.S., ***Degeneration Regeneration,***1949, 1977; Nutritional Development, 5235 Gulf Blvd., St. Petersburg, FL 33706. Also available from Price Pottenger Nutritional Foundation, P.O. Box 2614, La Mesa, CA 92041.

Passwater, Richard, Ph.D., **EPA-Marine Lipids**, Keats Publishing, Inc., New Canaan, CT, 1982.

Passwater, Richard, Ph.D., ***Evening Primrose Oil,*** 1981.

Passwater, Richard A., Ph.D. & Elmer M. Cranton, M.D., ***Trace Elements, Hair Analysis and Nutrition,*** Keats Publishing Inc., New Canaan, CT, 1983.

Pauling, Linus, Ph.D., ***How to Live Longer and Feel Better,*** W.H. Freeman & Co., New York, NY, 1986.

Pearson, Durk & Sandy Shaw, ***Life Extension, a Practical Scientific Approach,*** Warner Books, 75 Rockefeller Plaza, New York, NY 10019, 1982.

Russell-Manning, Betsy, ***Candida, Silver Fillings and the Immune System,*** Greensward Press, 1600 Larkin, #104, San Francisco, CA 94109, (415) 928-4142. 1985. Cancer Control Society, 2043 N. Berendo St., Los Angeles, CA 90027; Other books: ***How Safe are Mercury Fillings?; Wheatgrass Juice, Gift of Life.***

Sampsidis, N., M.S., ***Homogenized!,*** Sunflower Publishing Co., P.O. Box 25, Glenwood Landing, New York, NY 11547.

Schechter, Steven R., N.D., with Tom Monte, ***Fighting Radiation with Foods, Herbs & Vitamins,*** East West Health Books, 17 Station Street, Brookline, MA 02146, 1988.

Shannon, Sara, ***Diet for the Atomic Age,*** Avery Publishing Group, Inc., Wayne, NJ, 1987.

Sies, H. & A. Wendel, ***Functions of Glutathione,*** Springer-Verlag, NY, 1978.

Waldbott, George L., M.D., A.W. Burgstahler, Ph.D., L.H. McKinney, Ph.D., ***Fluoridation, the Great Dilemma,*** Coronado Press, Inc., Lawrence, KS, 1978.

Yiamouyiannis, John, ***Fluoride, the Aging Factor,*** 6439 Taggart Road, Delaware, OH 43015, 1983.

ORDER FORM

RADIATION PROTECTION MANUAL

3rd Edition

________ Copies at $9.95 each = ________
(includes shipping/handling)

________ Copies at $8.95 each = ________
(orders of 5 or more, includes shipping/handling)

Tax (California, 7.25%) = ________

Quarterly Newsletter ($40) = ________
(1st issue, Feb. 1990)

List of publications and tapes (please check) ________

TOTAL ENCLOSED = ________

Make check payable to: **Lita Lee, 2061 Hampton Ave.,**

Redwood City, CA 94061

Name (print clearly please) ________________________

Address (please give street address if possible, for UPS shipment):

____

Street Apt.

____

City State

____

Zip